Everyday Investigations for General Chemistry

Sally Solomon
Susan Rutkowsky
Charles Boritz

Drexel University
Department of Chemistry

WILEY

John Wiley & Sons, Inc.

ACQUISITIONS EDITOR	Nicholas Ferrari
PROJECT EDITOR	Joan Kalkut
PRODUCTION EDITOR	Elizabeth Swain
COVER DESIGNER	Jeof Vita
EXECUTIVE MARKETING MANAGER	Amanda Wainer

Library of Congress Cataloging-in-Publication Data

Solomon, Sally.
  Everyday investigations for general chemistry / Sally Solomon, Susan Rutkowsky,
Charles Boritz.
      p. cm.
   ISBN 978-0-470-08510-3 (pbk.)
   1. Chemistry, Physical and theoretical--Laboratory manuals. I. Rutkowsky, Susan. II.
Boritz, Charles. III. Title.
   QD457.S65 2008
   542--dc22

                              2008008337

10 9 8 7 6 5 4 3 2

Preface

PURPOSE OF THE BOOK

Introductory chemistry courses were once confined to a systematic treatment of elements and their compounds. Over the years the content has become more general with the introduction of physical chemistry topics and removal of much of the factual descriptive material. More changes have followed: space is now devoted to materials science as well as organic and biological chemistry. Practical applications help students relate concepts in chemistry to their surroundings. Traditional methods for teaching chemistry may be accompanied, or even replaced by, on line presentation.

This new laboratory manual, intended to accompany any mainstream general chemistry course, consists of experiments that can be completed using only chemicals found in consumer products. The manual is an ideal resource for courses emphasizing *green chemistry* in which the use of hazardous materials is reduced or eliminated altogether. Many of the experiments requiring simple equipment and glassware can be performed at remote sites providing laboratory experience for use with on-line or long distance learning courses.

The advantages of using accessible materials in chemistry laboratory are considerable. Students *can* reinforce lecture discussions while working with familiar materials. For instructors, assembling the chemicals required for a lab course can be accomplished with limited budgets and without access to a chemical company. Problems with safety and waste disposal are significantly reduced.

FEATURES

This manual is unique and different from all others in the market in that *all* of the experiments it contains can be done with chemicals and reagents found in drugstores, supermarkets, or convenience stores.

Using products: When possible, experiments are simply modified to utilize household chemicals. For instance, the titration of $HCl_{(aq)}$ with $NaOH_{(aq)}$ is performed with muriatic acid, a lye solution, and curcumin, an indicator extracted from turmeric. When substitutes are not available, new experiments have been designed. The qualitative analysis of household compounds using a flowchart replaces the analysis of common cations and anions. Available chemicals are used as starting materials to prepare new ones. Among these are the synthesis of ammonium chloride from household cleaner and muriatic acid and the fragrant oil of ethyl salicylate from aspirin tablets, muriatic acid, and ethanol.

Guided inquiry: One part of each experiment in the manual requires students to develop and carry out their own procedure for a given task. These guided inquiry sections also provide practical experience in reporting results with properly labeled plots, tables and diagrams.

Safety: Every experiment in the manual includes a safety section, which rates the toxicity, flammability, and exposure from 0 (low) to 3 (high) of all chemicals used.

Prelab: Questions are intended to practice skills needed for the experiment

Postlab: Questions following each lab require students to think about the experiment and the results they've obtained.

ORGANIZATION

The 27 experiments in the manual are arranged in the same order as topics in typical mainstream general chemistry textbooks. Experiments requiring a combination of concepts are placed at the end of the book.

The experimental write-ups include all the standard sections --- prelab, purpose, introduction, procedure, data/result tables, and postlab questions.

Guided inquiry is reserved for the final part of each experiment so that students will have already practiced the necessary techniques. This also provides instructors with the option of omitting these sections if faced with time constraints.

Computer-interfaced procedures that require colorimeter or gas pressure sensor probes are accompanied by traditional methods using spectrometers and barometers.

SUPPLEMENTS

The *Instructor's Guide* contains:

• A list of all of the equipment and chemicals needed for each experiment
• Instructions for preparing all solutions and reagents
• Useful ideas and information that expand upon the introductory material
• Suggestions for performing experiments at a remote site
• Typical results including completed data sheets and plots
• Answers to pre-lab and post-lab questions
• References

Acknowledgments

There are many people without whom this text could never have been produced. We wish to thank the following reviewers for their constructive comments and for the many positive remarks that helped provide us with the inspiration we needed to finish this project.

Linda Baker	Georgia Perimeter College
Craig Bayse	Old Dominion University
Robert Berger	Indiana University—Purdue University, Fort Wayne
Kenneth Capps	Central Florida Community College
Rajeev Dabke	Columbus State University
Jay Deville	Henderson State University
Michael Denniston	Georgia Perimeter College
John Francis	Columbus State Community College
Jason Hofstein	Siena College
Michael Rathmill	Community College of Philadelphia

We are particularly grateful to our project editor, Joan Kalkut, who was patient and always helpful --- ready to provide immediate feedback whenever we had a question. We were continually impressed by her ability to make the right suggestion about whatever it was that we were trying to do and to keep us on schedule (almost).

Contents

I. Chemicals

What makes this lab manual unique is that the chemicals used to perform all of the experiments it contains can be found in products readily available in supermarkets, hardware stores or convenience stores. Many compounds and aqueous solutions can be taken directly from their containers. Others are extracted from mixtures or synthesized starting with materials on hand.   Names, formulas, properties and sources of chemicals are listed in Tables A.2 – A.6 in Appendix A.

A. USED "AS IS"

Many products consist of one compound whose purity is, at worst, equivalent to technical grade materials (> 90%) from a chemical company. Generic brands of sucrose, sodium chloride and baking soda differ from their name-brand counterparts in packaging only. In other cases, there are differences.   The Arm and Hammer® brand of washing soda, a product used as a detergent booster, contains more than 98% sodium carbonate, whereas generic brands may have less than 75%.

Among products available in solution are aqueous acetic acid (vinegar) and sodium hypochlorite (bleach). Tincture of iodine contains iodine, potassium iodide and alcohol.

B. EXTRACTED

In extracting substances, a solvent is added to a solid mixture (such as naturally occurring spices, leaves, roots and barks) in order to separate one substance. For example, anthocyanin dye, obtained by treating red cabbage with isopropyl alcohol (Experiment 8), is an excellent universal indicator used throughout this text. Colors for pH ranges are listed in Table I.1.   Extracting curcumin (another pH indicator) from the spice, turmeric, is the subject of Experiment 9.

Color of Anthocyanin	pH Range
red	1-3
pink	4
violet-pink	5-6
violet	7
blue	8
blue-green	9
green-blue	10-12
yellow green	>12

Table I.1 pH Ranges for Anthocyanin

C. SYNTHESIZED

The variety of compounds can be increased by using accessible materials to synthesize new ones. Ammonium chloride, for instance, is not readily available, but can easily be prepared by mixing muriatic acid and household ammonia (Experiment 11):

$$HCl(aq) + NH_3(aq) \longrightarrow NH_4Cl(s)$$

Reactions such as this are the basis of original experiments in which the synthesis of a compound is followed by a study of its properties. Other substances synthesized include two copper pigments malachite and verdigris (Experiment 24) and salicylic acid and ethyl salicylate (Experiment 25),

Leftover product can be stored in properly labeled containers for use in other experiments. A permanent marker should be used on label paper that has a good adhesive. Labels must include the following:

- *Name*: spell out the chemical name.
- *Concentration*: if the chemical is in solution, indicate the molarity or mass%.
- *Hazards*: using words rather than symbols.
- *Date*: year prepared

For example. the label for solid NH_4Cl prepared in 2007 *must* have the following:

Ammonium chloride
Slightly hazardous upon skin
contact, ingestion, or inhalation.
2007

Adding other information such as the formula and molar mass is optional.

II. Equipment, Glassware, and Laboratory Technique

Diagrams of equipment and glassware used throughout this manual of experiments are shown. Essential laboratory techniques are discussed.

A. SMALL EQUIPMENT

Test tube racks and wash bottles are handy to have. The Bunsen burner is required only occasionally in performing the experiments in this manual.

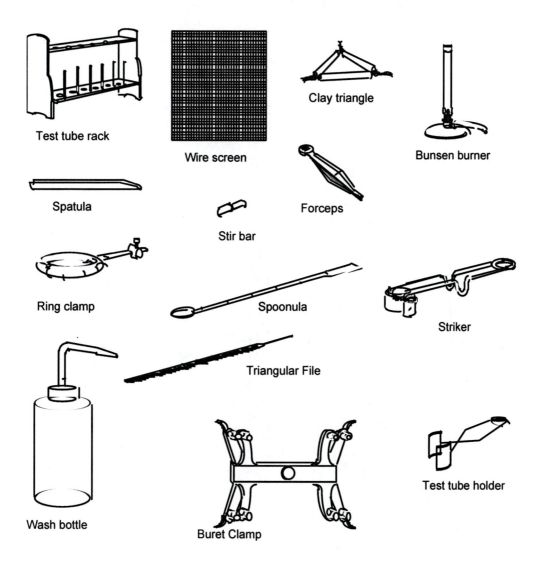

Figure II.1 Small Equipment

B. GLASSWARE

Common glassware used in the laboratory is pictured.

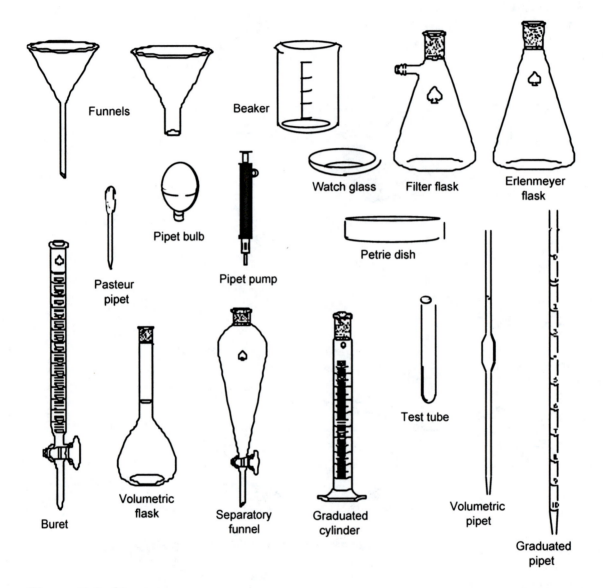

Figure II.2 Glassware

Of the glassware shown in the diagram, the graduated cylinder, pipets, buret, beaker and flasks are all involved in the delivery of volumes of liquid.

C. MEASURING VOLUME

The accuracy of the volumetric devices varies widely from the crude measurement provided by a beaker to the hundredth of a milliliter delivered by a volumetric pipet.

1. Meniscus

If poured into clean glassware, most solutions assume a concave upper surface with a well-defined meniscus. It is always the *bottom* of this meniscus that must be matched with the marking(s) on the glass wall. While checking the meniscus, you may find it helpful to use a piece of white paper as a backdrop.

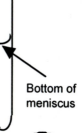

Bottom of meniscus

2. Graduated Cylinders

Graduated cylinders are calibrated for pouring out measured quantities of liquids. That means that a cylinder filled up, say, to the 20 mL mark contains just a little more liquid than 20 mL, since the last couple of drops will never come out (they remain smeared all over the wall of the cylinder).

3. Volumetric Flasks

Volumetric flasks contain a well-defined volume of liquid when filled to the mark on the neck. They are most commonly used for preparing and diluting solutions, a routine task performed in the chemistry laboratory.

4. Burets

Before filling a buret, it must be washed with tap water, then with the solution to be used. To *wash with water*, the stopcock is closed and the buret is filled all the way up from the top with tap water. A few milliliters are allowed to flow through the tip and the rest of the water is poured out. This is repeated twice. To *wash with solution*, the stopcock is closed and some solution is poured from a clean beaker to fill the buret about 1/4th full. After a couple of milliliters flows out to rinse the tip of the buret, the stopcock is closed and the remaining contents poured from the top into a waste beaker. This is done twice.

To *fill the buret*, 50-100 mL of solution is poured into a clean beaker. The buret is closed and mounted on a ring stand with a buret clamp. Using a funnel, the solution is poured carefully into the closed buret until the meniscus is about 1 or 2 cm above the '0' mark. Air bubbles that cling to the inside of the buret can be removed by tapping gently. The meniscus is lowered precisely to the '0' mark, checking the tip for trapped air.

5. Pipets/Pipet Pump

The *transfer* or *volumetric* pipet (the one with a single upper mark) is designed to deliver a fixed volume of liquid. The *graduated* pipet may be used to deliver any desired amount.

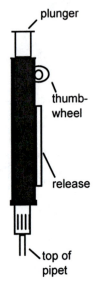

plunger

thumb-
wheel

release

top of
pipet

To fill either pipet using the pipet pump shown on the left, the plunger is pushed all the way down, and the pump is attached to the top of the pipet. The lower tip of the pipet is submerged into the solution without touching the bottom of the container. By rotating the thumbwheel on the pipet pump liquid is drawn into the pipet. The liquid should not go up into the pump. When the meniscus is 2-3 cm *higher* than the upper ('0') mark, the tip of the pipet is removed from the liquid. The meniscus will drop. If it drops below the upper ('0') mark, resubmerge the tip and draw more liquid into the pipet. If the meniscus drops to a point above the upper ('0') mark, use the thumbwheel to lower the meniscus to the mark. After filling the pipet to the upper mark, the release is used to transfer the measured amount.

In order to measure out a volume with a graduated pipet, the meniscus is lowered from the upper to the lower mark. The volume indicated on a volumetric pipet is measured out by letting go all the liquid below the mark without forcing out the last drops. The lower tip of the pipet should be touched to the wall of the vessel holding the solution, so that no drop remains hanging on the tip.

D. MEASURING MASS

When we speak of the weight of a substance, what we really mean is its mass. However, the familiar terms "weighing" and "weighed", rather than "massing" or "massed" will be used throughout this text.

Most of the experiments in this text can be done with top loading balances weighing to the nearest 0.01 g compared to analytical balances, the ones with sliding doors (nearest 0.001 or 0.0001 g). *Never* overload balances. Maximum load without damaging the top loading balance is 400 ± 0.01 g, and for the more sensitive analytical balance, 300 ± 0.001 g.

Taring is zeroing a container prior to adding a sample. To do this, the balance is reset by pressing the "TARE" key on the front face. Depending upon the sensitivity of the balance, this should give a digital display of either 0.00 g or 0.000 g. Containers must be placed on the balance before taring. Then the mass of the sample in the container is the displayed number.

Solid objects can be weighed directly. Powders require weighing papers or boats. Liquids are weighed in small glassware such as vials or in covered weighing bottles. The sample being weighed should be near the center of the balance pan. The balance pan must be cleaned after use.

E. FILTERING

In many of the experiments in this manual, filtration is done by gravity using an Erlenmeyer flask and a funnel fitted with filter paper. The paper should be 5-10 mm below the top of the funnel. There are two ways to prepare the paper, flat or fluted. A piece of round filter paper, folded in half then in quarters, is opened so that the paper cone is in contact with the funnel. Fluted paper can be purchased or can be made by folding round filter paper into halves then quarters followed by bringing each edge into the center to produce new folds. Care should be taken not to crease the folds too tightly at the center, which could weaken the fluted filter and cause it to break during the filtering process.

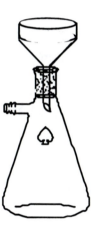

For vacuum filtering a filter flask is fitted with a Büchner funnel, a porcelain funnel with a flat perforated bottom. The side tube of the filter flask is connected to a vacuum. A trap bottle is positioned between the filter flask and the vacuum source. Before starting the filtration, a flat circle of filter paper placed in the funnel is moistened with solvent and suction is applied.

F. DRYING

If an oven is not available for drying, a wide shallow Pyrex dish and a hot plate will suffice for removing water from compounds. Care must be taken not to use a setting too high which could result in cracking the dish and spewing out its contents. As always, safety glasses must be worn during drying and all other activities.

G. COMPUTER-INTERFACING

The popular Vernier system includes a variety of probes that can be connected to an interface panel then to a laptop computer. In the setup below a conductance probe is shown.

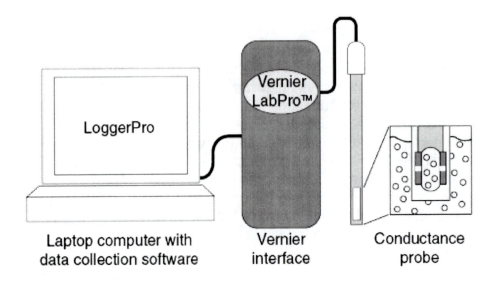

| Laptop computer with data collection software | Vernier interface | Conductance probe |

Among other probes used in experiments in this manual are the gas pressure sensor, colorimeter and the radiation monitor. Traditional procedures with barometers and spectrometers are included along with the computer-interfaced methods that require gas pressure sensors (Experiment 12) and colorimeter probes (Experiment 20).

III. Reporting Data and Results

Take data as you work. Don't worry about how neat it looks. For many of the experiments you will need to plot data and attach each graph to your lab report. For the first part (or parts) of each experiment the data and results sections already include prepared tables for entering data.

In most of the experiments there is one section that will ask you to design and carry out your own procedure for a particular task. These sections will require that you prepare your own result sheets which, depending upon the experiment, may include properly labeled plots, tables and diagrams. Instructions for plotting graphs (using Excel) and creating tables are given in the following sections.

A. GENERAL GUIDELINES FOR GRAPHING

A sample plot will be based on a set of data that was collected to test Boyle's Law (See Table III.1). The experimenter varied the volume(V) of air using a syringe and recorded the pressure (P) at each volume. The pressure could be read to ±0.4 mm Hg and the volume set to ±0.05 mL.

Table III.1 Pressure Volume Data for Air

P (mmHg)	V (mL)	1/V (mL^{-1})
763.9	10.0	0.100
703.9	11.0	0.0909
646.5	12.0	0.0833
600.5	13.0	0.0769
557.1	14.0	0.0714
520.9	15.0	0.0667
489.4	16.0	0.0625
462.2	17.0	0.0588
436.7	18.0	0.0556
417.1	19.0	0.0526
396.2	20.0	0.0500

1. What to plot?
According to Boyle's Law the pressure of a gas, P, is inversely proportional to the volume, V, and directly proportional to its reciprocal, 1/V. Plotting P vs. V would give a hyperbola and plotting P vs. 1/V, a straight line. Linear plots are preferred for several reasons. Deviations are readily seen by eye whereas deciding if a curve that resembles a hyperbola is actually a hyperbola is not obvious. Linear plots can also be extrapolated (extended) into regions, which are experimentally not accessible. Thus, we will plot P vs. 1/V.

2. Which variable should be on the x-axis?
The *independent variable*, the one controlled in the experiment, is the x-axis. In this experiment the experimenter systematically alters the volume (V) and measures the value of the pressure (P) at each volume. The x-axis is the volume variable or 1/V in mL^{-1} and the y-axis is P, mm Hg.

3. To use Excel

Using a plotting program such as Excel, enter the data you wish to plot in two columns, with the horizontal scale (x-axis) points in the first or left-hand column and the vertical scale (y-axis) points in the right-hand column. Select the cells of the spreadsheet containing the values to be plotted. Click on the **ChartWizard** which will open to display a gallery of Chart types. Choose XY (Scatter) and the Chart sub-type version with points only. Enter the Chart title (Boyle's Law for our sample) concisely worded, to tell what the graph is about. Label each axis with the variable and its unit.
For our plot, $1/V$ (mL^{-1}) was entered for the x axis and P (mm Hg) for the y-axis.
When you Finish with Chart Wizard you can reformat the chart as desired. For example, clicking on the shaded chart area and choosing No Fill replaces the gray with white.

4. Scale

Select an axis and adjust the Scale by changing the settings for Maximum and Minimum. The lower limit of each axis need not be zero. Choosing zero for the lower limit of P, for example, would cause all the points to be crammed into a small area of the graph. In general, try to have the plot fill as much as possible of the chart area.

5. To add error bars

Follow these steps:
• Click a point in the data series to which you want to add error bars.
• On the Format menu, click Selected Data Series.
• On the X Error Bars tab or the Y Error Bars tab, select the options you want.

Here we can use the Fixed value of 0.4 mmHg for the Y error bar. The error for $1/V$ changes with the volume, V, which can be read to ±0.05 mL. For the X error bar, we use the maximum error in mL^{-1} which is 0.001, calculated for 10 mL, the smallest volume used. Select Display error bars (for both the x and y-axes). When the error bars are too small to show on the graph, the points remain as the original data marker symbols. In the finished plot on page 11, notice that the Y error bars are not seen and the X error bar is small.

6. Trendline

A *trendline* as shown in the finished graph can be added to make the best straight line. From the Chart menu select Add Trendline. Choose Linear. If desired, click on Options and check the box for Display equation on chart. Excel will display an equation for a straight line giving its slope. Modify the equation as needed to include the correct number of significant figures.

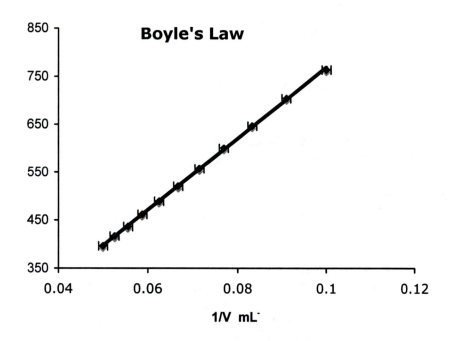

Boyle's Law

If a plot is done on graph paper, you can fill in the trendline by hand. Do not "connect the dots". To find the slope, choose two points on the line and divide the difference in the y values by the difference in x values.

$$Slope = \frac{y_2 - y_1}{x_2 - x_1}$$

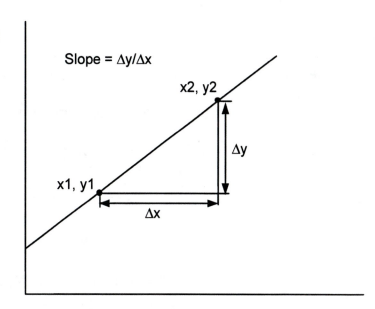

7. The point 0,0
The point 0,0 is sometimes a data point. If it is, do not forget to include this in your spreadsheet data list so that it will be included on your chart. For this sample plot, there is no 0,0 data point.

8. More than one curve
Where two or more curves are plotted on one sheet for comparison, the curves should be differentiated by using different types of lines (dotted, dashed), or by different colors. The plotted points for the curves can be circles, squares, triangles, etc.

B. GENERAL GUIDELINES FOR CREATING TABLES

The format and structure of tables should be simple and well designed for ease and clarity of reading. Try to keep tables as uncluttered as possible.

1. Titles
Tables should have simple concise titles. If an entry is the same for an entire column, it should be part of the title instead.

2. Column Headings
The first row of the table containing physical quantities should consist of the quantity and its unit.

3. Numerical Values (and powers of 10)
If a power of ten follows each entry in a column of data, it should be in the header as the reciprocal. For example, an entry '2.0' under the heading 10^3 (unit) means that the value of the quantity is 0.0020; an entry '2.0' under the heading 10^{-3} (unit) means that its value is 2000. Placing the power of ten in the column heading also removes ambiguity about the number of significant figures in a quantity. In the example given, there are two significant figures.

The tables below list values of conductance for NaCl solutions. Notice that Table III.3 is well designed and clear compared to Table III.2.

Table III.2 Conductance of various concentrations of NaCl solutions

Molarity	Conductance	Temperature
1.72×10^{-3}	210	25
2.59×10^{-3}	315	25
3.45×10^{-3}	415	25

Table III.3 Conductance of $NaCl_{(aq)}$ at 25°C

Molarity (mol/L x 10^3)	Conductance (μS)
1.72	210
2.59	315
3.45	415

IV. Safety

Every experiment in the manual includes a section on safety where hazards are reported for all chemicals used. The data is assembled from information provided by Flinn Scientific (http://www.flinnsci.com/search_MSDS.asp) and from Jay Young's CLIPS (Chemical Laboratory Information Profiles) series published in the Journal of Chemical Education.

Hazards graded from 0 (low) to 3 (high) are assigned in the areas of toxicity, flammability and exposure. The table below is taken from an experiment using the solvents acetone and methanol.

Chemical	Toxicity	Flammability	Exposure
acetone (B.P. 56°C)	1	3	1
methanol (B.P. 65°C)	3	3	1

Hazards for all chemicals used in this manual are tabulated in this way. Special precautions accompany the tables. For the one above, it would be pointed out that methanol can be absorbed through the skin.   Waste disposal suggestions are included as well.

Experiment 1
Density: Identifying Solids and Liquids

_____

Name_____ Date_____ Section_____

1. From your personal observation, decide which substance in the pairs below has a higher density. Give a reason for each choice.

a) gasoline or water?

b) water or ice?

c) fresh water or seawater?

2. Before 1982 U.S. pennies were made primarily of copper (density is 8.9 g/mL) and after 1982, mainly zinc (density is 7.1 g/mL). Given that the volume of a penny is 0.35 mL, compare the mass of a 1972 penny to the mass of one dated 2002. Assume that the pennies are either 100% copper or 100% zinc.

3. One so-called "cube" of sugar, from a box of "sugar cube dots" used to sweeten coffee or tea, has a mass of 2.3 g and the dimensions 1.1 cm x 1.1 cm x 1.2 cm. Find its density.

*Instructor's Signature*_____

Experiment 1
Density: Identifying Solids and Liquids

PURPOSE
Densities will be used to aid in identifying various liquids and solids.

INTRODUCTION
In this experiment, the mass (m) and volume (V) of a substance will be measured, and its density (D) calculated. The units are g/mL (the same as g/cm^3).

$$D = \frac{m}{V}$$

Note that density is an intrinsic property, which is the same regardless of the quantity of the substance. Therefore, density can be used to aid in the identification of a substance.

Liquids
Densities can be useful in partial identification of a liquid, that is, in ruling out some possibilities. Table 1.1 lists common liquids and their densities at 20°C and 25°C. Replacing an atom in a formula with one of lower atomic weight (say C for O) usually decreases the density of a liquid. For example, the density of gasoline, a mixture of hydrocarbons, is less than 0.7 g/mL.

Table 1.1 Densities of Liquids

Liquid	Formula	D (g/mL) (20°C)	D (g/mL) (25°C)
water	H_2O	1.000	0.997
ethyl acetate	$C_4H_8O_2$	0.9006	0.8945
methanol	CH_4O	0.7915	0.7866
ethanol	C_2H_6O	0.7895	0.7852
isopropyl alcohol	C_3H_9O	0.7854	0.7812
gasoline	$*C_nH_{2n+2}$	0.66 – 0.69	

* the major components of gasoline are hydrocarbons called alkanes, C_nH_{2n+2}, where n ranges from 5-12.

The density measurement is made by weighing a given volume of the liquid. Using covered weighing bottles helps to reduce errors resulting from evaporation of low-boiling liquids.

Solids

Solids can also be partially identified by measuring their densities. Densities of common solids are listed in Table 1.2 including different kinds of glass. About 90% of the glass manufactured is soda lime silica used for windows and light bulbs. Lab glassware, headlamps and cookware are borosilicate, which resists temperature changes.

Table 1.2 Densities of Solids

Solid	D (g/mL)
copper	8.93
brass	8.5
steel	7.8
calcium carbonate/ marble, chalk	2.7 - 2.8
soda-lime silica/ windows, light bulbs	2.5
borosilicate/ labware, cookware, head lamps	2.1 - 2.3
beeswax (C, H, O)	0.96-0.97
paraffin wax (C, H)	0.87-0.91

Densities of many solids can be determined by simply finding the mass and volume of a sample. In this experiment some of the solids will be large enough to use displacement of water or calculation from dimensions to find their volumes. Other methods, such as flotation, are needed for measuring densities of small objects with irregular shapes.

In the flotation method or "sink and float" a sample is suspended in a liquid mixture. The mixture contains one liquid with a density a little larger than the object to be measured. Then a second liquid, much less dense than the first but entirely soluble in it, is added in small portions until the solid is suspended. The density of the mixture is the density of the solid object. Flotation is commonly used for measuring glass fragments found at the scene of a crime. However, one solvent needed is $CHBr_3$, bromoform (D 2.9 g/mL), a dangerous sedative that is not readily available. All glass samples in this experiment will be large enough to determine their volumes by calculation or displacement of water.

APPARATUS

The mass in grams is measured using a top loading balance (nearest 0.01 g). The balance is reset by pressing the "ZERO" or "TARE" key on the front face, giving a digital display of 0.00 g.

⚠CAUTION: Do not overload balances. The maximum load that can be weighed without damaging the top loader is 400 g, a little less than one pound.

SAFETY

Wear safety glasses throughout this experiment. Solvents must be handled with care. Methanol is absorbed through the skin. Wear gloves throughout this experiment. Note that "absolute" alcohol, which is close to 100% ethanol, may contain traces of isopropyl alcohol, methanol or benzene, all of which are toxic.

Chemical	Toxicity	Flammability	Exposure
ethyl acetate	1	3	1
methanol	3	3	1
ethanol	2	3	1
isopropyl alcohol	3	3	1
calcium carbonate	0	0	1
soda lime	1	0	2
borosilicate	0	0	0
beeswax	0	0	0
paraffin wax	0	1	0

0 is low hazard, 3 is high hazard

PROCEDURE

Part A: Density of Liquids

1. Record room temperature and sample code on the data sheet.

2. Since most of the liquid unknowns evaporate readily, a covered container is recommended. Either a weighing bottle with lid or a small beaker (30-mL or 50-mL) covered with Parafilm will do. Place the container and cover on the balance, press "Zero" or "Tare", then remove them from the balance.

3. Using a volumetric pipet (right) with a pump or bulb, pipet 10.00 mL of sample into the weighing bottle. Cover immediately, then place the weighing bottle on the balance. Record the mass.
The volume indicated on the pipet is measured out by letting go all the liquid below the mark. You must not force out the last drops. Just touch the lower tip of the pipet to the wall of the container, so that no drop remains hanging on the tip.

⚠**CAUTION:** Never pipet by mouth.

4. Calculate the density and then use the liquid densities in Table 1.1 to identify the unknown.

Part B: Density of Solids

Two different methods are given for identifying solid unknowns. Choose the best one to use for each of your unknown samples. Give a reason for each decision.

Method 1 Density by Displacement of Water

1. Weigh the sample to the nearest 0.01 g.

2. Select a graduated cylinder into which the sample will fit. The sample must have a volume of at least 10 mL; if needed use two samples. Before placing the sample in the cylinder, be sure it contains enough water to cover the sample (30-35 mL). Record the volume of water. Then place your sample into the graduated cylinder. Record the new volume. Remove bubbles attached to the sample by shaking or tapping the cylinder.

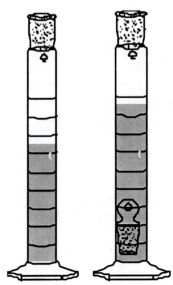

3. Calculate the density (g/mL) of the sample and record. Because of the error in the volume measurement, the density may be off by as much as 0.2 to 0.3 g/mL. With this in mind, use the data in Table 1.2 to identify your unknown.

Method 2 Density by Flotation

1. Pour 25 mL water into a 50-mL graduated cylinder.

2. Add the sample solid. If it floats the density of the unknown must be less than 1.

3. Choose another solvent that is miscible with water and has a lower density.

4. Add the solvent until the object is suspended.

5. Remove a 10.00 mL sample of the liquid mixture and find its density. Identify the solid from the table of densities.

Data and Results (Density)

Name_____ Date_____ Section_____

Part A: Density of Liquids

Temperature_____ °C

Sample Code	Volume Liquid (mL)	Mass Liquid (g)	Measured Density (g/mL)	Identity of Liquid & Tabulated Density*

* If needed, estimate the tabulated density for your sample. For example, suppose the temperature of your sample is 22.5°C. You can assume that the tabulated density is halfway between the recorded values at 20°C and 25°C.

Part B: Density of Solids

Method 1 Displacement of Water

Sample Code	Mass (g)	Vol. H_2O in Grad (mL)	Vol. H_2O + Solid (mL)	Vol. of Solid (mL)	Density (g/mL)	Identity of Solid

Reason for choosing displacement of water:

*Instructor's Signature*_____

Data and Results (Density)

Part B: Density of Solids

Method 2 Density by Flotation

Sample Code _____

Volume water _____ mL

Solvent added _____

Temperature _____ °C

Density of solvent added _____ g/mL

Volume solvent required for sample to sink _____ mL

Density liquid mixture:

           Mass of 10.00 mL _____ g

           Density of mixture _____ g/mL

Identity of solid _____

Reason for choosing flotation:

Questions

1. Why must the volume of the solid to be determined by water displacement be at least 10 mL? Explain.

2. What is the advantage of using a weighing bottle rather than an open vessel for weighing liquids?

3. In measuring the density of a low-boiling liquid, a 10.00-mL sample is transferred to an open beaker, then weighed. Would you expect the calculated density to be too high or too low?

Experiment 2
Beer's Law, Copper Ammonium Complex

Name_____ Date_____ Section_____

1. Give formulas for the compounds below. How do the formulas of the colored compounds differ from the white ones?
a) sodium bicarbonate (white)

b) calcium chloride (white)

c) magnesium sulfate heptahydrate (white)

d) copper(II) sulfate pentahydrate (blue)

e) iron(III) chloride hexahydrate (yellow)

f) iron(II) chloride tetrahydrate (light green)

2. Give the range (in nm) of the visible portion of the electromagnetic radiation spectrum. Indicate the location of red and violet.

3. Four components of a spectrometer include the following:
      Detector; prism (or diffraction grating); sample; source
a) What is the role of the prism or diffraction grating?
b) Make a block diagram of a spectrometer showing the four parts given.

Instructor's Signature _____

Experiment 2
Beer's Law, Copper Ammonium Complex

PURPOSE
Using Beer's Law the concentration of the deep blue copper ammonium complex, $Cu(NH_3)_4^{2+}$, will be found from the amount of light absorbed by its aqueous solution.

INTRODUCTION
One very convenient way to find the concentration of a colored solution is from its visible spectrum.

Spectroscopy
A spectrum is a recording of the wavelengths absorbed by a sample. Colored compounds, such as the copper ammonium complex used in this experiment absorb in the visible (350 to 700 nm). The diagram below includes the major components of a common single-beam spectrometer:

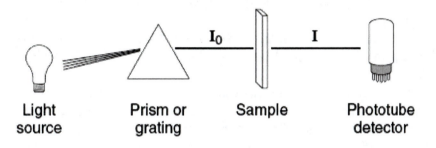

| Light source | Prism or grating | Sample | Phototube detector |

In the Spectronic-20, the light source is a tungsten bulb. The light is separated into wavelengths with a diffraction grating and the detector is a phototube. The ratio of the intensity of the emerging light (I) and the intensity of the incident light (I_0) is the percent transmittance, %T:

$$\%T = \frac{I}{I_o}$$

Absorbance, A, is the logarithm of 100 divided by %T; a scale reading of 50% T corresponds to log 2 or 0.30 on the A scale:

$$A = \log\frac{100}{\%T}$$

and,

$$0.30 = \log\frac{100}{50}$$

Since it is easier to read % T rather than A, the %T values are recorded and converted to A. The spectrum is plotted with the wavelength on the x-axis and the absorbance on the y-axis.

Beer's Law
According to Beer's Law, the absorbance (A) of a solution is proportional both to the concentration (c) of the solute responsible for the color and to the path length (b) of solution through which the light passes.

$$A = \varepsilon \, b \, c$$

The path length is the diameter of the tube used to hold the sample. When the concentration is in molarity and the path length in cm, 'ε' is the *molar absorptivity* with the units $cm^{-1} \, mol^{-1} \, L$. (Molar absorptivity has replaced the term extinction coefficient.) The value of 'ε' depends on the wavelength of light, the solvent and the kind of solute used. Deviations from Beer's Law may be severe for concentrations greater than 0.1 M.

Copper Ammonium Complex,
The copper ammonium complex or tetraamminecopper(II) complex, is prepared by adding aqueous ammonia to the hydrated sulfate of tetraaquocopper(II), $Cu(H_2O)_4SO_4 \cdot H_2O$ (usually written $CuSO_4 \cdot 5H_2O$). Aqueous $Cu(H_2O)_4^{2+}$ reacts with ammonia to form a copper ammonium ion:

$$Cu(H_2O)_4^{2+} + 4NH_3 \rightarrow Cu(NH_3)_4^{2+} + 4H_2O$$

The major complex ion produced when the ammonia concentration is greater than 0.01 M is the intensely blue colored tetraamminecopper(II) ion, $Cu(NH_3)_4^{2+}$.

The hydrated copper sulfate used in this experiment, which is present in root killer products, controls a broad spectrum of pests, including fungi and algae. The active component is the copper(II) or cupric ion which binds to sulfide groups in enzymes of organisms, thus denaturing their proteins and leading to cell leakage. Very low concentrations of copper sulfate are also used to control parasites and algae in fish tanks.

APPARATUS
The spectrum is taken using a Spectronic-20 or similar device. To use the instrument refer to the operating instructions.

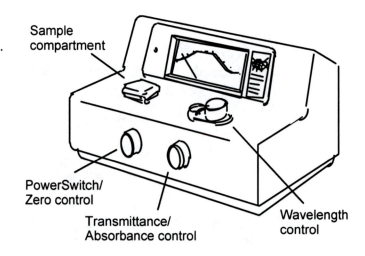

Sample compartment

PowerSwitch/ Zero control

Transmittance/ Absorbance control

Wavelength control

SAFETY
Wear safety glasses and gloves throughout this experiment. Copper sulfate stains the skin.

Chemical	Toxicity	Flammability	Exposure
0.10 M $CuSO_4 \cdot 5H_2O$	1	0	1
1 M ammonia	2	0	2

0 is low hazard, 3 is high hazard

PROCEDURE

Part A: Taking the Spectrum of 0.0100 M Copper Ammonium Complex
Solutions of 0.100 M $Cu(H_2O)_4^{2+}$ and 1 M ammonia will be available.

1. To prepare a stock solution of 0.0100 M $Cu(NH_3)_4^{2+}$, pipet 10.00 mL of 0.100 M $Cu(H_2O)_4^{2+}$ solution into a 100-mL volumetric flask. Dilute to the mark with 1 M NH_3. The stock solution is the most concentrated solution to be used in this experiment and will be used to take a spectrum and to prepare dilutions for other parts of the experiment.

2. Fill one cuvet with the 0.0100 M $Cu(NH_3)_4^{2+}$, solution (the sample) and another with 1 M ammonia (the blank).

3. Take the visible spectrum from 450 nm to 700 nm by measuring the % transmittance (Spectronic-20) every 10 nm, except near the maximum around 620 nm, where you should record %T every 5 nm.

4. Convert % transmittance to absorbance. Plot A vs. wavelength, by drawing a smooth curve through the points. Record the peak absorbance at λ_{max} which should be near 620 nm.

Part B: Measuring Absorbance of Diluted Solutions of 0.0100 M $Cu(NH_3)_4^{2+}$
To verify Beer's Law you will now prepare dilutions of a stock solution to measure their absorbances at the maximum wavelength of the spectrum curve found in Part A.

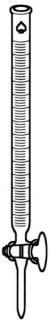

1. Pour about 50 mL of the stock 0.0100 M $Cu(NH_3)_4^{2+}$ into a clean 150-mL beaker. Fill a clean buret with the contents of the beaker.

2. Using two burets (one for stock 0.0100 M $Cu(NH_3)_4^{2+}$, another for 1 M ammonia), deliver 10.0 mL of the stock solution into a clean 250-mL beaker, then add 5.0 mL of 1 M ammonia to dilute the solution. Mix well. Using the same technique, prepare three additional dilutions by adding other volumes of water to three more 10-mL portions of the stock solution.

  10.0 mL stock + 10.0 mL 1 M ammonia
  10.0 mL stock + 15.0 mL 1 M ammonia
  10.0 mL stock + 20.0 mL 1 M ammonia

Calculate the concentrations of these diluted solutions and record them on the data sheet.

3. Measure the % transmittance of each diluted solution at the peak wavelength observed in Part A. Rinse the cuvette with each new solution before reading its % transmittance.

4. Convert % transmittance to absorbance. Plot the absorbance vs. the concentration. Be sure to make the points in your plot reflect the experimental error. The plot should produce a straight line. From the slope, determine 'ε', the Beer's Law constant, molar absorptivity. Submit the plot with your Data and Results. Remember that the point 0,0 is also a data point. Using a spreadsheet program such as Microsoft® Excel, find the equation of the trend line, y = mx + b, where 'm' is the slope.

5. Record the code number/letter of your unknown. Measure the % transmittance of the sample at the peak wavelength used for the known samples. Use the calculated value of 'ε' to find the concentration of your sample or use the Beer's Law plot to find the concentration corresponding to the measured absorbance.

Part C: Using $Cu(H_2O)_4^{2+}$ to Determine Concentration of Cu^{2+}
Evaluate the use of $Cu(H_2O)_4^{2+}$ compared to $Cu(NH_3)_4^{2+}$ to measure the concentration of an unknown Cu^{2+} solution. Include a procedure along with data and plots needed to support your conclusion.

Data and Results (Beer's Law)

Name_____ Date_____ Section_____

Part A: Taking the Spectrum of 0.0100 M Copper Ammonium Complex

Concentration of stock solution of $Cu(NH_3)_4^{2+}$ _____ M

λ (nm)	%T	A	λ (nm)	%T	A
450			600		
460			605		
470			610		
480			615		
490			620		
500			625		
510			630		
520			635		
530			640		
540			650		
550			660		
560			670		
570			680		
580			690		
590			700		

Peak wavelength where A is maximum (λ_{max}) _____ nm

A at λ_{max} _____

Instructor's Signature _____

Data and Results (Beer's Law)

Part B: Measuring Absorbance of Diluted Solutions of 0.0100 M $Cu(NH_3)_4^{2+}$

1 M NH_3 Added (mL)	Molarity	%Transmittance	Absorbance
0			
5			
10			
15			
20			

Path length _____ cm

Molar absorptivity, ε _____ $M^{-1}cm^{-1}$

(Attach absorbance vs. concentration plot)

Code number/letter of the Unknown _____

Absorbance of the Unknown at the Peak Wavelength _____

Concentration of the Unknown _____ M

Part C: Using $Cu(H_2O)_4^{2+}$ to Determine Concentration of Cu^{2+}

Questions

1. How could you use Beer's Law to find the concentration of a solution that is about 0.2 M?

2. Why is it best to use the wavelength at the absorption maximum for measuring concentration?

3. When determining copper(II) using Beer' Law, why is the hydrated copper(II) converted to the copper ammonium complex ?

4. In plotting a spectrum, why is wavelength the x-axis?

Prelaboratory Exercise

Experiment 3
Analysis of Food Dyes using Visible Spectroscopy and Paper Chromatography

Name_____ Date_____ Section_____

1. Of the spectra shown below where the absorbance, A, is plotted vs. wavelength in nm, which one is of: a) a green solution?   b) a pink solution?
Explain how you made your decision.

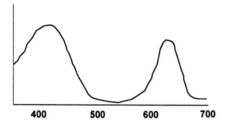
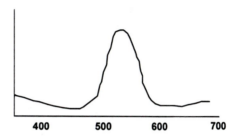

2. At approximately what wavelength would you expect to find the absorption maximum of an intensely blue solution?

3. Chromatography separates a mixture by distributing its components between a stationary phase and a mobile phase. Is this a physical process or a chemical process? Make an educated guess about the origin of the term chromatography.

Instructor's Signature _____

Experiment 3
Analysis of Food Dyes using Visible Spectroscopy and Paper Chromatography

PURPOSE
Components in food dyes and their concentrations will be compared for different brands of food dye.

INTRODUCTION
The dyes present in food colors are studied using two analytical techniques---visible spectroscopy and paper chromatography.

Food Dyes.
The five organic compounds shown below (along with Blue No. 2 and Green No. 3) are the only coloring additives certifiable for general food use in the United States. The largest group of permitted dyes are azo (-N=N-) compounds including the two yellows and Red No. 40. The green food dye is made by combining blue and yellow dyes.

Allura Red
FD&C Red No. 40
λmax = 507 nm

Erythrosine
FD&C Red No. 3
λmax = 530 nm

Sunset Yellow
FD&C Yellow No. 6
λmax = 480 nm

Tartrazine
FD&C Yellow No. 5
λmax = 426 nm

Brilliant Blue
FD&C Blue No. 1
λmax = 630 nm

Colors and Wavelength
When white light, which contains all of the colors in the visible spectrum, is passed through a colored sample, the sample will absorb certain wavelengths and transmit those that are not absorbed. The transmitted wavelengths will be seen as a color. For example, yellow dyes absorb blue (426 and 480 nm) and appear yellow; red dyes absorb in the green (500 and 530 nm) and look red. Blue dye absorbs yellow light (630 nm) and transmits blue. This is the principle upon which the laundry product called

"bluing" works. White fabrics tend to turn slightly yellow after extended use. Bluing contains a dye such as Prussian Blue, $Fe_4[Fe(CN)_6]_3$, which is suspended in water. When bluing is added to laundry rinse water it adds a trace of blue color to fabrics making them appear whiter because the blue dye absorbs yellow. The use of bluing as a rinse for white hair is based on the same idea.

Visible Spectrum

A visible spectrometer will be used to measure the spectrum of dye samples, that is wavelengths of light they absorb. The spectrum is a plot of A (y-axis) vs. wavelength in nm (x-axis). Absorbance or transmittance can be read from the instrument meter. However, it is easier to read % T rather than A, so record the %T values and convert them to A. Absorbance, A, is the logarithm of 100 divided by %T:

$$A = \frac{100}{\%T}$$

For example, if %T = 50, then A = log 100/50 = log 2 = 0.30. Notice that 50% T corresponds to 0.30 on the Absorbance scale.

Paper Chromatography

In paper chromatography, components of a mixture will separate according to how strongly they adsorb on a stationary phase (filter paper) versus how readily they dissolve in a mobile phase (solvent mixture or developing solution). The mobile phase travels up the stationary phase, carrying the samples with it. Procedures specify the composition of the solvent mixture, the paper used (often Whatman filter paper) and the means by which the spots are visualized.

In this experiment the mixtures are food colors, consisting of the five permitted FD&C food dyes. The stationary phase is filter paper and the mobile phase is 0.1% NaCl. The separated dye mixture will be clearly visible as a series of colored spots.
How far each dye moves up the paper depends on its solubility in the liquid versus how much is retained by the paper. The distance between the original spot, its final location, and the total distance traveled by the solvent itself is known as the R_f (retardation factor) value:

R_f = Distance moved by the spot/ Distance moved by solvent front

The table gives R_f values from use of Whatman #1 filter paper and 0.1% NaCl. The R_f's increase from 0.065 for Red No.3 which moves the least, to 0.95 for Blue No. 1, which moves the farthest.

Dye	Color	R_f
Blue No. 1	bright blue	0.95
Yellow No. 5	lemon yellow	0.58
Yellow No. 6	orange yellow	0.44
Red No. 40	orange red	0.27
Red No. 3	pink	0.065

APPARATUS

A visible spectrometer such as the Spectronic-20 is used to take the spectra of the dyes.

Spots of each food color are applied to filter paper, which is placed in a covered beaker with the developing solution.

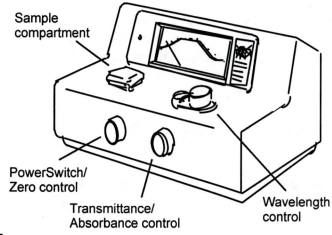

Sample compartment

PowerSwitch/ Zero control

Transmittance/ Absorbance control

Wavelength control

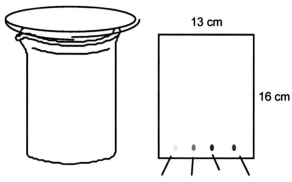

13 cm

16 cm

SAFETY

Wear safety glasses throughout this experiment.

Chemical	Toxicity	Flammability	Exposure
food dyes	0	0	0
0.1% NaCl (aq)	0	0	0

0 is low hazard, 3 is high hazard

PROCEDURE

Part A: Taking the Spectrum of a Food Dye

To save time and to avoid repeating similar procedures, each group can take the spectrum of one color, then share results with other groups.

1. Set up the Spectronic-20 according to operating instructions.

2. Dilute the dyes by adding one drop of food coloring from a Pasteur pipet to a 100-mL volumetric flask then diluting with water to the mark. This should place the maximum absorbance between 0.5 and 1.5.

3. Measure the % T every 10 nm between 400 nm and 700 nm, remembering to zero the instrument with a water blank at each new wavelength. After determining the approximate wavelength of maximum absorbance, you may want to "redo" the region around the peak wavelengths proceeding with 5 nm increments.
Keep the solutions. Later you may be able to see a difference among spectra of different brands by eye.

4. Convert % T to A and plot absorbance vs. wavelength, by drawing a smooth curve through the points. This is the visible spectrum of the stock solution of food dye.

5. Record the wavelength and peak height where the maximum absorbance is observed. Include major peaks. The larger the peak height the greater the concentration.

Part B: Paper Chromatography
Food colors are readily separated into their constituent dyes by using paper chromatography with dilute salt water, 0.1% NaCl (aq), as the mobile phase and Whatman # 1 filter paper as the stationary phase.

1. Take a section of filter paper measuring 13 cm by 16 cm. Dip the sharp end of a toothpick or a capillary tube into the concentrated food dye from the original container and carefully touch the toothpick or tube to the paper about 3 cm from the wide edge of the paper. Good results require that the spots be no larger than 2 mm in diameter once they have wet the paper. Allow the spots to dry.

2. Place the paper in a 600-mL beaker (spots near the bottom) and add 0.1% NaCl (aq) to a level just below the spots. Cover with a plate or watch glass and allow the chromatogram to develop for about 10-15 minutes. As the solvent system migrates up the paper, the dye compounds in the spot are carried along at different speeds by the moving liquid. Be sure that the solvent front does *not* go all the way to the top of the paper.

3. When the solvent is near the top of the paper, remove the paper and immediately make a pencil mark to indicate how far the paper has become wet from the upward movement of the salt water. Record the distance measured from the origin.

4. Fill in the dye composition chart, listing each FD & C dye found in each color, and measured R_f values.

Part C: Comparing Other Brands of Food Dyes
Materials needed for this section are provided. Outline procedures, record data and results using tables and plots as needed.

Compare different brands of food dyes. Choose one color, the same one used for the brand chosen in Part B. Share results with other groups.

Data and Results (Food Colors)

Name_____ Date_____ Section_____

Part A: Taking the Spectrum of Food Dye

Color_____   Brand _____

Brand:			Brand:		
λ (nm)	%T	A	λ (nm)	%T	A
350			520		
360			530		
370			540		
380			550		
390			560		
400			570		
410			580		
420			590		
430			600		
440			610		
450			620		
460			630		
470			640		
480			650		
490			660		
500			680		
510			700		

Brand _____ drops/100 mL

Color	λ_{max} (nm)	A at λ_{max}

Instructor's Signature _____

Data and Results (Food Colors)

Part B: Paper Chromatography

Stationary Phase _____   Mobile Phase _____

Distance traveled by solvent _____cm

Dye	Distance Traveled by Dye (cm)	R_f, Measured
Red No. 40		
Red No. 3		
Blue No. 1		
Yellow No. 5		
Yellow No. 6		

Dye composition (enter dyes present in each color for the brand you tested)

Brand _____

Red	Yellow	Blue	Green

Data and Results (Food Colors)

Part C: Comparing Other Brands of Food Dyes

Questions

1. Compare concentrations of dyes of different brands. Comment on cost and concentration.

2. Why was a Pasteur pipet used to measure *one* drop of food dye?

3. The sample cell used for most visible spectroscopy is a glass test tube or cuvet. Could this cell be used to take a spectrum of a sample that absorbs in the ultraviolet, say, around 280 nm? What information is needed to answer this question?

4. Some types of chromatography can be used to separate mixtures, collecting each component. Why would this be difficult to do using paper chromatography?

Prelaboratory Exercise

Experiment 4
Determining the Formula of a Hydrate

Name________________________________ Date_____________ Section_____

1. Cobalt (II) chloride is used for detecting moisture. Its anhydrous form is blue and the hydrated form is pink. If one mole of the pink hydrate contains 108 g water, what is its formula?

2. Anhydrous calcium chloride, used as a drying agent, can be prepared by heating its hydrated salt to drive off the water. If heating 200 g of hydrated calcium chloride produces 151 g anhydrous salt, what was the formula of the hydrate?

3. The capacity of a drying agent is the amount of water it can remove per gram. The four hydrates below form when the drying agents potassium carbonate, calcium sulfate, sodium sulfate, or calcium chloride are used at room temperature. For which one is the capacity: Very high? High? Medium? Low?
a) potassium carbonate dihydrate
b) calcium sulfate hemihydrate (hemi is ½)
c) sodium sulfate decahydrate
d) calcium chloride hexahydrate.

*Instructor's Signature*________________________________

Experiment 4
Determining the Formula of a Hydrate

PURPOSE
Thermal dehydration will be used to find formulas for both fully hydrated salts and their intermediate hydrates.

INTRODUCTION
Salts that crystallize from water along with water molecules are called hydrates. Most of the water molecules (sometimes all of them) will bond to the metal atom to form a hydrated ion. For instance, aluminum chloride crystallizes from solution along with six water molecules. In this case all six waters are bonded to the Al atom. The formula is preferably written as $Al(H_2O)_6Cl_3$, the chloride of the hydrated aluminum ion, hexaaquo aluminum. However, formulas of hydrated salts are more often simplified with all waters of hydration attached to the end of the formula as in $AlCl_3 \cdot 6H_2O$. The compound $AlCl_3 \cdot 6H_2O$, also called aluminum chlorhydrate, is the main active ingredients in many antiperspirants. The aluminum ions are taken into the cells that line the sweat glands at the opening of the epidermis, the top layer of the skin. When the aluminum ions are drawn into the cells, water passes in with them. As more water flows in, the cells begin to swell, squeezing the ducts closed so that sweat can't get out.

Not all waters in hydrates are necessarily the same. For example, in the structure of hydrated copper(II) sulfate, four of the water molecules bond with the copper cation to produce, $Cu(H_2O)_4^{2+}$, called the tetraaquocopper(II) ion. The 5th water is weakly bonded to $Cu(H_2O)_4SO_4$ and is more easily removed than the other four. The formula of the hydrated sulfate of tetraaquocopper(II) should be written as $Cu(H_2O)_4SO_4 \cdot H_2O$ and should be called tetraaquocopper(II)sulfate hydrate. But this is rarely done. Instead the formula is normally written as $CuSO_4 \cdot 5H_2O$, and is called copper(II) sulfate pentahydrate.

Hydration and Shelf Life of Products
The bound waters contained in a product such as a pharmaceutical tablet can influence its shelf life as well as its properties and physical appearance. The temperature(s) at which dehydration occurs provides valuable information about the stability of the hydrates. Copper sulfate is used to test blood for anemia and has been used as an ingredient in copper reduction tablets that test for the presence of glucose in urine. Magnesium sulfate or Epsom Salt is a cathartic and an anti-inflammatory agent. The evolution of water from materials such as the hydrates of copper sulfate and magnesium sulfate has been thoroughly studied. The dehydration of copper sulfate pentahydrate takes place in three stages as shown in the plot.

The waters are given off in the sequence 2-2-1. At 87°C, the first two waters are removed, the hydrogen-bonded lattice water and one of the four waters bonded to the Cu ion. Two more waters come off at 116°C and the last one requires a temperature over 230°C.

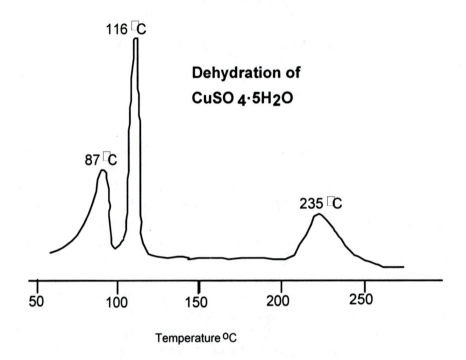

116 °C

Dehydration of
CuSO$_4$·5H$_2$O

87 °C

235 °C

50   100   150   200   250

Temperature °C

The conversion of the hydrate of magnesium sulfate to the monohydrate takes place in multiple steps below 150°C. At 200°C anhydrous magnesium sulfate is produced.

APPARATUS

A sample of hydrated salt is placed in a crucible on top of a clay triangle that is supported on a ring clamp for heating with a Bunsen burner.

The hotplate and Petri dish setup below provides the more controlled heating required to achieve the temperatures needed for partial dehydration.

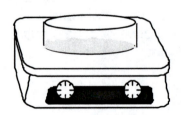

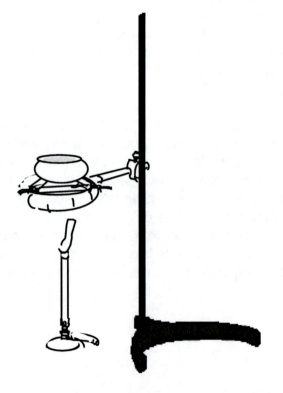

SAFETY
Wear gloves throughout this experiment. Copper sulfate stains the skin.

Chemical	Toxicity	Flammability	Exposure
$CuSO_4 \cdot 5H_2O$	2	0	2
$MgSO_4 \cdot 7H_2O$	0	0	1
$Na_2B_4O_7$	1	0	1

0 is low hazard, 3 is high hazard

PROCEDURE
Salts can be completely dehydrated using a Bunsen burner in Part A. In Part B, partial dehydration is possible with more controlled heating from a hot plate. In Part C, an attempt is made to remove the waters of hydration one or two at a time.

Part A: Complete Dehydration

1. Clean a crucible and rinse thoroughly with distilled water.

2. Place the crucible on a clay triangle that is supported on a tripod. Heat for about five minutes with a Bunsen flame to dry the crucible. Cool the crucible back to room temperature. Weigh and record the mass to the nearest 0.01 g.

3. Place 3-4 g hydrated copper sulfate crystals in the crucible. Heat for about 15 minutes to drive off the water. (The copper sulfate will decolorize.) Cool to room temperature. Weigh the crucible.

⚠**CAUTION:** Do not allow the crucible to become red hot. Overheating can cause the copper sulfate to decompose forming the dark-colored oxide.

4. Reheat the crucible and contents for another 5 minutes. Cool and reweigh. The amounts should agree within 0.03 g. If not, reheat for another 5 minutes, cool, reweigh and check.

5. Find the formula of the hydrated copper sulfate. Assume complete dehydration. Repeat for Epsom Salt.

Part B: Partial Dehydration

1. Place 3-4 g hydrated copper sulfate in a Petri dish. Using a setting of a little above 100°C on a hotplate, heat for about 10 minutes, cool and weigh. Reheat, cool and reweigh until the amounts agree within 0.03 g.

⚠**CAUTION:** If the temperature is set too high, the Petri dish can break.

2. Find the formula of the partially dehydrated copper sulfate: calculate the moles of water driven off per mole of hydrated salt used. Use the formula of the hydrated salt found in Part A.

3. Repeat for Epsom Salt.

Part C: Formation of $CuSO_4 \cdot 3H_2O$
Determine whether or not it is feasible to produce the trihydrate of copper sulfate using the equipment available for this experiment. Describe your procedure and record all data.

Data and Results (Water of Hydration)

Name_____ Date_____ Section_____

Part A: Complete Dehydration
Assume that all the water was driven off.

	Hydrated Copper Sulfate	Hydrated Magnesium Sulfate
mass empty crucible (g)		
mass crucible + hydrated salt (g)		
mass crucible + salt after heating		
first (g)		
second (g)		
third (g)		
if needed, fourth (g)		
mass water lost (g)		
moles water lost (mol)		
mass dehydrated salt (g)		
molar mass dehydrated salt (g/mol)		
moles dehydrated salt (mol)		
mol H_2O lost /mol dehydrated salt		
Formula of hydrated salt		

Other observations:

*Instructor's Signature*_____

Data and Results (Water of Hydration)

Part B: Partial Dehydration

	Hydrated Copper Sulfate	Hydrated Magnesium Sulfate
mass empty Petri dish (g)		
mass dish + hydrated salt (g)		
mass dish + salt after heating		
first (g)		
second (g)		
third (g)		
if needed, fourth (g)		
mass water lost (g)		
moles water lost (mol)		
mass hydrated salt (g)		
molar mass hydrated salt (g/mol)		
moles hydrated salt (mol)		
mol H_2O lost /mol hydrated salt		
Formula of partly dehydrated salt		

Part C: Formation of $CuSO_4 \cdot 3H_2O$

Questions

1. Why is it necessary to know the formula of fully hydrated copper(II) sulfate in order to find the formula of its intermediate hydrate?

2. Upon prolonged vigorous heating of 2.49 g (0.0100 mol) of copper sulfate pentahydrate, 0.795 g of a black residue forms along with another product that vaporizes and is lost. What is the black compound? Give a balanced chemical equation for the degradation reaction.

3. Why must a hot plate (or oven) be used in Parts B and C, rather than a flame?

Experiment 5
Quantitative Determination of P in Plant Food

Name_____ Date_____ Section_____

1. Find the mass of chloride present in 5.00 g sodium chloride.

2. The salt in Number 1 is to be analyzed for chloride by precipitation and weighing of silver chloride.
a) Why is it important to make the sodium chloride solution using distilled water?
b) What mass of silver chloride would be obtained?
c) What property of silver chloride makes this determination possible?

3. Suppose a sample containing magnesium oxide is to be analyzed by converting it to insoluble $Mg_2P_2O_7$. If 1.35 g of the magnesium pyrophosphate is produced, what amount of MgO was present?

Instructor's Signature _____

Experiment 5
Quantitative Determination of P in Plant Food

PURPOSE
The percentage of phosphorus (as P_2O_5) in a sample of commercial plant food will be determined by precipitation of $MgNH_4PO_4 \cdot 6H_2O$.

INTRODUCTION
A description of plant foods and the analytical method used here to determine their phosphorus content follows.

Plant Food

Like animals, plants require essential nutrients supplied by fertilizers in a form that can be utilized by the plant. The three most likely to be lacking in soil are the primary nutrients, nitrogen, phosphorus and potassium. Three numbers on the labels of plant foods indicate respectively the percentage of nitrogen (N), percentage of phosphorus (as P_2O_5) and the percentage of potassium (as K_2O). For instance, the common all-purpose water-soluble plant food product now containing 24% N, 8% P_2O_5, and 16% K_2O is labeled 24-8-16, which means that it is 24% N, 3.5 % P, and 13.3 % K. (This is a new product that was recently reformulated by Miracle-Gro to replace the original all-purpose product, 30-15-30.) The P content is quoted as P_2O_5 content even though there is no P_2O_5 in the plant food. Likewise the plant foods contain no K_2O. The source of N, P, and K in plant foods is listed in Table 5.1, and the other components in Table 5.2.

Table 5.1 Source of N, P, and K in Plant Foods

Name	Formula
Urea	$(NH_2)_2CO$
Ammonium phosphates	$(NH_4)_2HPO_4$, $NH_4H_2PO_4$
Potassium chloride (muriate of potassium)	KCl

Table 5.2 Other Components

Name	Formula
Boric Acid	H_3BO_3
Copper Sulfate	$CuSO_4$
Chelated Iron EDTA	Fe_2EDTA
Manganese sulfate	$MnSO_4$
Zinc sulfate	$ZnSO_4$
Dye	Unknown (proprietary)

Analytical Method

The gravimetric determination of phosphorus is based on the precipitation of insoluble magnesium ammonium phosphate hexahydrate from a solution that contains acid phosphate ions, ammonium ions, and magnesium ions. The precipitate forms upon slow neutralization with ammonia of an acidic solution of a phosphate containing sample. An excess of ammonium ion (derived from the plant food which contains ammonium phosphates) encourages the formation of $MgNH_4PO_4 \cdot 6H_2O$ through the common ion effect.

$$5H_2O(l) + HPO_4^{2-}(aq) + NH_4^+(aq) + Mg^{2+}(aq) + OH^-(aq) \rightleftharpoons MgNH_4PO_4 \cdot 6H_2O(s)$$

The $MgNH_4PO_4 \cdot 6H_2O$ does not form in acidic solution since the concentration of phosphate is reduced by conversion to acid phosphate:

$$PO_4^{3-}(aq) + H_3O^+(aq) \rightleftharpoons HPO_4^{2-}(aq) + H_2O\ (l)$$

The hydroxide needed for the neutralization must be provided by a weak base such as ammonia, since a strong base like NaOH would cause the precipitation of $Mg(OH)_2$ and other undesirable compounds.

In classic schemes the precipitate would be converted by ignition to the pyrophosphate, $Mg_2P_2O_7$, and weighed:

$$MgNH_4PO_4 \cdot 6H_2O(s) \longrightarrow Mg_2P_2O_7(s) + 13\ H_2O\ (g) + 2\ NH_3(g)$$

Fortunately, the $MgNH_4PO_4 \cdot 6H_2O$ is sufficiently stable at room temperature to be dried and weighed thus avoiding the ignition and the difficulties associated with it.

APPARATUS

No special apparatus is needed. This experiment was originally designed to use only equipment and chemicals readily available to most people. The procedures can be done with crude balances in any building which has running water---or in an academic laboratory.

SAFETY

Wear safety glasses throughout this experiment.

Chemical	Toxicity	Flammability	Exposure
$MgSO_4 \cdot 7H_2O$	0	0	1
ammonia (household)	2	0	2
70% isopropyl alcohol	3	2	0

0 is low hazard, 3 is high hazard

PROCEDURE
In the first week the P is precipitated as $MgNH_4PO_4 \cdot 6H_2O$.

Part A: Preparing the Reagents

1. You may need as much as 180 mL of 10% magnesium sulfate heptahydrate (Epsom salts) solution per unknown sample. Prepare enough for two runs (in case something goes wrong) by placing 20 g Epsom Salts in a large beaker or flask and mixing with tap water to a volume of approximately 200 mL. This should dissolve easily. You can stir to assist the solution process. The exact concentration is not important.

2. The other reagents, rubbing alcohol, and ammonia, will be used in Part B.

Part B: Precipitation of $MgNH_4PO_4 \cdot 6H_2O$

1. Choose a sample of plant food and record the number code on your data sheet.

2. Prepare a sample that weighs about 10 grams. The mass need not be exactly 10.0 g, but you must record it to the nearest 0.1 g and you must be sure that the mass is at least 10 grams or a little greater so that you will have 3 significant figures to work with.

3. Mix the plant food with about 120 mL of tap water in a 250-mL Erlenmeyer flask. Swirl until dissolved. Filter to remove any undissolved material, using filter paper and a funnel. Filter into a 1-Liter Erlenmeyer flask. Wash the residue remaining in the filter paper with 10 to 20 mL of tap water. Plant food is advertised to be entirely soluble, however there may be some residue. Estimate the weight of this, or at least set an upper limit to it.

4. You will be adding 140 to 180 mL of the magnesium sulfate solution to your dissolved sample. Pour the solution from a graduated cylinder so that you can record the amount added to the nearest 5 mL. Swirl to mix. The solution is acidic at this point and must be neutralized to produce the insoluble $MgNH_4PO_4 \cdot 6H_2O$.

5. Place in a clean container about 180 mL of ammonia solution taken from a bottle of household ammonia. Add the ammonia from a clean graduated cylinder, about 25 mL at a time to your mixture of plant food water and magnesium solution. (You may need more than 100 mL, and as much as 170 mL; it is best to use too much than not enough). A white precipitate will form. The solution must be alkaline (basic) for precipitation to be complete. Test by using pH paper provided. If the solution is still acidic keep adding ammonia. Record the amount of ammonia you use to the nearest 5 mL. Allow the mixture to stand for a few minutes before filtering.
You may notice foaming because some ammonia cleaning products contain detergent. This will not interfere with anything unless the product contains phosphorus. The label on any of the brands should say "contains no phosphorus".

6. Prepare a filtering setup using a flask and funnel.
If the filter paper breaks, don't start over!   Place everything mixture, filter paper and all in a beaker and refilter using a different kind of filter paper!

7. Wash with two (or more as needed) 30-mL portions of rubbing alcohol. Do this by adding each portion to the flask in which you did the precipitation to gather any remaining material, then pour through the solid in the funnel. Record the total amount of alcohol used for washing on the data sheet.   As soon as the alcohol has filtered through it is possible to remove the filter paper with great care, spread it out on a paper-towel placed on top of a paper plate. Smooth the solid paste out for easy drying. Be sure to recover anything that sticks to a spatula used for smoothing. Allow to dry until next week.

Part C: Finding % P (2nd Week)

1. Weigh the product by carefully scraping it off the paper. You *cannot* measure the mass of precipitate by first weighing the empty filter paper and then the filter paper plus precipitate. The filter paper itself always absorbs an indeterminate amount of water.

2. Calculate % P and % P_2O_5.

3. Ask your instructor for the package of plant food from which your unknown sample was taken. Record the "actual" amount of phosphorus, listed on the label as "%P_2O_5" Compare with your experimental value.

Data and Results (P Determination)

Name_____ Date_____ Section_____

Part B: Precipitation of $MgNH_4PO_4 \cdot 6H_2O$

1. Mass of plant food sample _____ g

2. Mass of water insoluble residue (estimated): Less than _____ g

3. Volume of magnesium sulfate solution _____ mL

4. Volume of ammonia _____ mL

Part C: Finding % P *

1. Mass of $MgNH_4PO_4 \cdot 6H_2O$ _____ g

2. Mass of plant food sample _____ g (See B.1)

3. Mass of P in unknown sample _____ g

4. Mass of P_2O_5 in unknown sample _____ g

5. % of P_2O_5 in the unknown sample: _____ (Use only 2 significant figures)

6. Brand/type of plant food _____

% P_2O_5 (See label) _____

7. Error (in % P as P_2O_5) _____
If your answer falls in the correct range your error is said to be zero since the actual amount is known so crudely. Otherwise, estimate your error to the nearest 0.5%.
*Another approach is to convert from g to mol $MgNH_4PO_4 \cdot 6H_2O$, then convert to mol P, to mol P_2O_5, and then to g P_2O_5.

Instructor's Signature _____

Questions

1. Why is the solution of dissolved plant food acidic?

2. In the presence of potassium ion, another precipitate could form in which potassium replaces ammonium.
a) Give the formula for this precipitate.
b) How would this influence your % P result (high answer or low answer)?
c) Would the error from sodium contamination be as serious? Why or why not?

3. Precipitation with NaOH(aq) will produce contaminating precipitates such as Mg(OH)$_2$. How would this affect the % P result?

4. Give the nutrient composition of a fertilizer containing only urea, (NH$_2$)$_2$CO.

Prelaboratory Exercise

Experiment 6
Preparation of Solutions

Name_____ Date_____ Section_____

1. What is the difference between molarity and molality? Define each one.

2. A 5.00% saline solution is prepared by mixing 5.00 g NaCl with 95.0 g H_2O
a) Find the molality.
b) If the density of this solution is 1.04 g/mL, find the molarity
c) Compare the molarity and molality of this solution. Why are they numerically similar?

3. The absorbance (A) of cobalt chloride, $CoCl_2 \cdot 6H_2O$, solution is 0.10 at 510 nm. The path length (b) is 1.0 cm and the molar absorptivity (ε) is 4.8 $M^{-1}cm^{-1}$. Using Beer's Law (A = ε x b x c) find the concentration in molarity (c) of the solution.

Instructor's Signature _____

Experiment 6
Preparation of Solutions

PURPOSE
Solutions will be prepared using the concentration units of molarity and molality.

INTRODUCTION
Preparing solutions is one of the most fundamental tasks performed in the laboratory.

Molarity: By far, the most common concentration unit used by chemists is molarity. Molarity (*M*) is defined as the number of moles (*n*) of solute dissolved in exactly one liter of solution.

$$\text{Molarity(M)} = \frac{n(\text{solute})}{L(\text{solvent})}$$

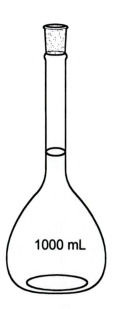

1000 mL

A measured amount of solute needed for a particular concentration is placed in a volumetric flask and solvent is added up to the mark on its neck. It is not necessary to use 1-L volumetric flasks such as the one shown here. They come in sizes as small as a few milliliters to as large as several liters.

Molality: Molality units are used in calculating changes in boiling point and freezing point that a solvent undergoes when solute is added.   Molality (m) is the number of moles of solute dissolved in exactly one kilogram (1000 g) of solvent.

$$\text{Molality(m)} = \frac{n(\text{solute})}{kg(\text{solvent})}$$

To make a solution of a given molality two measurements must be made, the mass of the solute and the mass of the solvent.

APPARATUS
A spectrometer such as the Spectronic-20 can be used to test concentrations of colored solutions. In this experiment the wavelength will be set to 700 nm where copper sulfate pentahydrate absorbs. To test the concentration of the solution you prepare its absorbance can be compared to the known value for that concentration.

Comparing the measured conductivity of a solution to the known tabulated value can tell you if the concentration of the prepared solution is correct.

One system for measuring conductance is the Vernier LabPro® interface and conductivity probe which is immersed in the solution to be tested. The probe is plugged into the interface, which is connected to a computer or hand-held device such as a graphing calculator for data collection.

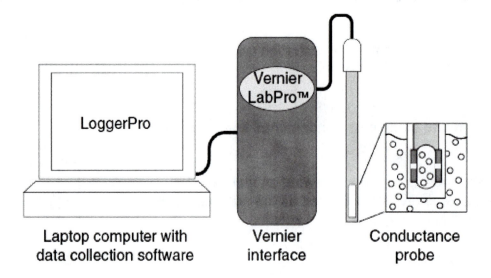

| Laptop computer with | Vernier | Conductance |
| data collection software | interface | probe |

SAFETY
Wear safety glasses and gloves throughout this experiment. Copper sulfate stains the skin. Sodium hydroxide is caustic.

Chemical	Toxicity	Flammability	Exposure
copper sulfate pentahydrate	2	0	1
sodium chloride	0	0	0
calcium chloride dihydrate	1	0	1
sodium hydroxide	3	0	3
magnesium sulfate heptahydrate	0	0	1

0 is low hazard, 3 is high hazard

This data is for the solid compounds. The hazard levels for aqueous solutions are lower for toxicity, flammability, and exposure.

PROCEDURE
Find the amount of solid solute needed for a solution of given volume and concentration. Record results of calculations in the tables provided. Prepare the solution, test to see if your concentration is correct, and find % error.
On the data/results page show calculations of the mass required for each of the solutions as well as the % error.

Part A: Preparation of 0.10 M Copper Sulfate Pentahydrate

1. Find the number of moles needed to make 50 mL (0.050 L) of 0.10 M solution using a 50-mL volumetric flask.

2. From the molar mass of copper sulfate pentahydrate, $CuSO_4 \cdot 5H_2O$ and the number of moles from Step 1, find the mass in grams of solute needed.

3. Weigh the mass of solute calculated from Step 2 and carefully transfer the blue solid to a clean 50-mL volumetric flask.

4. Add about half the volume of distilled water needed and swirl the flask. When most of the solid has dissolved add the rest of the water stopping below the mark on the flask. To add the remaining water use the water wash bottle. Insert the stopper and invert the flask a few times for uniform mixing. The *bottom* of the curved water surface, the "meniscus", should touch the mark on the neck of the volumetric flask.

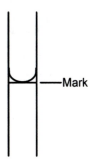
—Mark

5. Test to see if the concentration of your blue copper sulfate solution is correct by measuring its absorbance at a given wavelength. Fill a cuvet with your solution and place in the spectrometer which is set to a wavelength of 700 nm.   Record the absorbance, (A), which is directly related to concentration.

6. Find the % error. The absorbance at 700 nm should be 0.77.

Part B: Preparation of 0.20 m Sodium Chloride

1. Find the number of moles of NaCl that must be mixed with 100 g water to make a 0.20 m solution.

2. From the molar mass of NaCl and the number of moles from Step 1, find the number of grams of sodium chloride needed .

3. Weigh the amount of NaCl from Step 2 and transfer to a clean Erlenmeyer flask. Measure 100 mL water (100 g) using a graduated cylinder and pour into the 250-mL Erlenmeyer flask.   Swirl to dissolve.

4. Test the solution by measuring its conductance using the probe interfaced to the computer. Insert the probe in the solution so that the electrode surfaces are completely submerged in the liquid. Gently swirl the probe and wait about 5 to 10 seconds for a reading to appear.   A 0.20 m NaCl solution has a conductance of 18,500 µ S.   The unit for conductance, is the *siemen*, S, a very large unit. Conductance of aqueous solutions is measured in microsiemens, µS (micro is 1 millionth, 10^{-6}).

5. Find the % error.

Part C: Preparation of Additional Solutions

Prepare and test one (or more) solutions from the list below. Describe the method used for each one. Show calculations and tabulate results

Compounds and various sizes of volumetric flasks (the largest is 100 mL) are available. Conductance data is tabulated for testing the solutions below.

Solution	Conductance (µS) Tap Water	Conductance (µS) Distilled Water
0.050 M calcium chloride dihydrate	9614	9289
0.0010 M sodium hydroxide	706	267
0.010 M magnesium sulfate heptahydrate	2358	1928

Data and Results (Solutions)

Name_____ Date_____ Section_____

Part A: Preparation of 0.10 M Copper Sulfate Pentahydrate

To Prepare 50 mL 0.10 M $CuSO_4 \cdot 5H_2O$	Result
Moles needed (mol)	
Molar mass of $CuSO_4 \cdot 5H_2O$ (g/mol)	
Mass in grams needed (g)	
Absorbance at 700 nm, measured	
Absorbance at 700 nm, known	
% Error	

Calculations:

Instructor's Signature _____

Data and Results (Solutions)

Part B: Preparation of 0.20 m Sodium Chloride

Mass (g) of NaCl needed to mix with 100 g water to prepare a 0.20 m solution:

To Prepare 0.20 m NaCl using 100 g H$_2$O	Result
Moles needed (mol)	
Molar mass of NaCl (g/mol)	
Mass in grams needed (g)	
Conductance 0.20 m NaCl, measured	
Conductance 0.20 m NaCl, known	
% Error	

Calculations

Part C: Preparation of Additional Solutions
Tabulate quantities involved and % error. Show calculations.

Questions

1. Describe how you would prepare 100 mL 0.10 M NaOH.

2. Using 100 g water, how would you prepare 0.20 m NaOH?

3. Describe how you would prepare the following aqueous solutions. Include the amounts of substances and any special equipment you might need.
a) 1 L of 0.10 M KCl.
b) 250 mL of 1 M sucrose (molar mass 342)

Experiment 7
Analysis of Spinach Extract

_____

Name_____ Date_____ Section_____

1. What evidence do you have from everyday observation that room temperature water is unlikely to be effective in removing chlorophyll from leaves?

2. Why would you expect chlorophyll to absorb in the visible region of the spectrum? At what wavelength(s) would you expect the absorbance to be very low?

3. The success of a solvent used to separate a mixture by paper chromatography depends upon its polarity. For each of the following pairs of liquids, choose the one that is most polar. Give a reason for each choice
a) H_2O or C_5H_{12} (pentane).
b) C_3H_6O (acetone) or C_8H_{18} (octane)
c) H_2O or C_2H_6O (ethanol).

Instructor's Signature _____

Experiment 7
Analysis of Spinach Extract

PURPOSE
Visible spectroscopy will be used to estimate the ratio of chlorophyll a and b in an ethanol extract of spinach. A paper chromatography method using only solvents in consumer products will be designed for separating the two chlorophylls.

INTRODUCTION
Structure and properties of the pigments in spinach are described.

Extraction
In solid extraction, a liquid is used to remove one or more substances from a mixture of solids, often those that occur in nature. In this experiment the solid mixture is spinach, the liquid solvent is ethanol, CH_3CH_2OH, and the substance to be isolated is chlorophyll.

Chlorophylls and other compounds
Plants synthesize carbohydrates (sugars and starches) from CO_2 and H_2O through a complex set of reactions called photosynthesis. The first step is the absorption of sunlight by chlorophyll molecules in plants. There are two chlorophylls with closely related structures. The 'a' compound (left) has a methyl group, CH_3, where the 'b' form (right) has an oxygen-containing group, CHO. These are in bold in the formulas below.

chlorophyll a

chlorophyll b

Two other pigments present are β-carotene (top) and xanthophyll (bottom):

Paper Chromatography

In paper chromatography, components of a mixture will separate according to how strongly they adsorb on a stationary phase (filter paper) versus how readily they dissolve in a mobile phase (solvent mixture or developing solution). The mobile phase travels up the stationary phase, carrying the samples with it.

In this experiment the mixtures are plant pigments, consisting primarily of chlorophyll a and chlorophyll b. How far each pigment moves up the paper depends on its solubility in the liquid versus how much is retained by the paper. The distance between the original spot, its final location, and the total distance traveled by the solvent itself is known as the R_f (retardation factor) value:

$$R_f = \text{Distance moved by the spot/ Distance moved by solvent front}$$

An effective mobile phase is 90% petroleum ether and 10% acetone (See Table 7.1). Petroleum ether (PE), a mixture of the alkanes, pentane, C_5H_{12} and hexane, C_6H_{14}, is not readily available in consumer products. However, there are similar products. In Part C of this experiment you will be asked to try to find a substitute. The R_f values and some properties of the pigments present in spinach are summarized in Table 7.1.

Table 7.1

Compound	Solubility in Ethanol	Color	λ_{max} (nm)	R_f using 9 PE: 1 Acetone
chlorophyll a	very soluble	blue-green	429	0.65
chlorophyll b	very soluble	green to yellowish-green	453	0.45
β-carotene	slightly soluble	yellow-orange	445	0.71
xanthophyll	soluble	yellow	440	0.95

APPARATUS

A Spectronic-20 or similar instrument will be used to take the spectrum. Refer to the operating instructions.

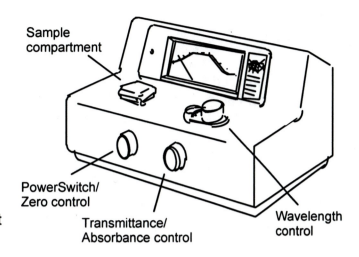

Solvents are tested for chromatographic separation of the two chlorophylls using a covered beaker and Whatman #1 filter paper. A successful separation is shown on the right side of the paper below.

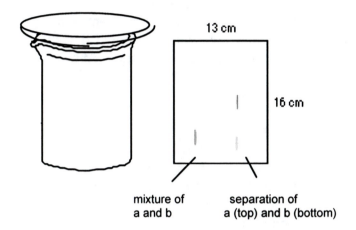

mixture of a and b

separation of a (top) and b (bottom)

SAFETY

Solvents must be handled with care. Methanol is absorbed through the skin. Wear safety glasses and gloves throughout this experiment. Note that "absolute" alcohol, which is close to 100% ethanol, may contain traces of isopropyl alcohol, methanol or benzene, all of which are toxic. Lighter fluid and charcoal lighter fluid are flammable and should be kept away from sparks and flames. They may be harmful upon skin contact, however only small amounts are used in this experiment.

Chemical	Toxicity	Flammability	Exposure
ethanol	2	3	1
acetone	1	3	1
5% acetic acid	0	0	0
methanol	3	3	1
ethyl acetate	1	3	1
ethylene glycol	2	1	2
isopropyl alcohol	3	3	1
lighter fluid	2	3	2
charcoal lighter fluid	2	3	2

0 is low hazard, 3 is high hazard

PROCEDURE

Part A: Extracting Chlorophyll from Spinach

1. Shred a medium-sized leaf of spinach and mix with 10 mL of ethanol in a 50-mL beaker. After approximately 5 minutes, the solution should have a green tint. If the solution looks yellow, the extraction is not finished.

2. Filter the mixture of leaves and ethanol into another 50-mL beaker, using a funnel with a fluted filter. Discard the spinach.

Part B: Spectrum of Chlorophylls

1. Fill one cuvet with ethanol (the blank) and another with the solution of chlorophylls (the sample). First, check the absorbance at 440 nm. The reading there, where both chlorophylls absorb to some extent, should be somewhere between 0.3 and 1.0. If the solution is too concentrated, pour the sample into a graduated cylinder and add an equal amount of alcohol to dilute the solution by half. Check the absorbance at 440 again. Dilute further if needed. Record the dilution needed.

2. Set the wavelength dial to 400 nm and measure the % transmittance. Continue to measure the % T every 10 nm between 400 nm and 700 nm, remembering to adjust the instrument to full scale at each new wavelength. Take readings every 5 n.n between 420 and 480 to make it easier to observe the absorption due to both chlorophylls.

3. Convert % T to A. Plot absorbance vs. wavelength by drawing a smooth line through the points (do not "connect the dots").

4. Examine the spectrum to find evidence for presence of both chlorophyll a and b. Estimate the relative amounts of a and b in your sample.

Part C: Paper Chromatography of Ethanol Extract

An effective solvent mixture known to separate the colored pigments in spinach is 90% petroleum ether(PE) and 10% acetone. For R_f values see Table 7.1. However, the petroleum ether, a mixture of pentanes and hexanes, is not readily available in any consumer product. Using the solvents that are available, listed in Table 7.2 in order of increasing polarity (alkanes to water), try to separate chlorophyll a from b. In general, polarity of the solvent mixture is increased until separation occurs. Use one liquid or a mixture (but no more than three).

With a toothpick, place spots of the concentrated extract on a large sheet of filter paper. Place a small drop of the solvent chosen on each spot. The extract will consist mostly of chlorophylls. Record qualitative results for the solvents chosen. Test to see if any liquid or mixture produces visible separation, that is, two spots, that should be visible as blue green and yellowish green. Attach filter paper that shows any evidence of separation. Share your results with other students.

In the table below, molecular formulas are used for the alkanes. However, one molecular formula may represent many different structures. For instance, C_5H_{12}, pentane, has three isomers:

$$CH_3CH_2CH_2CH_2CH_3$$

$$CH_3CH_2CHCH_3$$
$$|$$
$$CH_3$$

$$CH_3$$
$$|$$
$$CH_3CCH_3$$
$$|$$
$$CH_3$$

pentane or *n*-pentane where *n* means *normal* or straight chain	2-methylbutane or isopentane where iso means a methyl group is branched off from the next to last C	2,2-dimethylpropane or neopentane, the most highly branched of the pentane isomers

The alkanes in household products are usually the straight chain isomers, the ones that are least reactive, and hence are most abundant in petroleum deposits. For example, hexane, C_6H_{14}, is mostly $CH_3CH_2CH_2CH_2CH_2CH_3$.

Table 7.2 Solvents in order of increasing polarity; alkanes to water

Name	Formula	Source
Alkanes	C_nH_{2n+2}	
Light		
pentane (n =5)	C_5H_{12}	
hexane (n = 6)	C_6H_{14}	lighter fluid
heptane (n =7)	C_7H_{16}	
Medium		
octane (n =8)	C_8H_{18}	charcoal lighter fluid
nonane (n =9)	C_9H_{20}	
decane (n =10)	$C_{10}H_{22}$	adhesive remover
Heavy		
undecane (n =11)	$C_{11}H_{24}$	adhesive remover
dodecane (n =12)	$C_{12}H_{26}$	
tridecane (n =13)	$C_{13}H_{28}$	lamp oil
Ethyl acetate	$CH_3-\underset{\overset{\|\|}{O}}{C}-OCH_2CH_3$	nail polish remover (non-acetone)
Acetone	$CH_3-\underset{\overset{\|\|}{O}}{C}-CH_3$	nail polish remover; solvent
Isopropyl alcohol	CH_3CHOH $\|$ CH_3	rubbing alcohol (91%)
Ethanol	CH_3CH_2OH	200 proof alcohol (100% ethanol)
Water	H_2O	tap or distilled water

Data and Results (Chlorophyll)

Name_____ Date_____ Section_____

Part B: Spectrum of Chlorophylls

Dilution(if required): Vol. Leaf Extract _____mL   Vol. Alcohol added: _____ mL

λ (nm)	%T	A	λ (nm)	%T	A
400			530		
410			540		
420			550		
425			560		
430			570		
435			580		
440			590		
445			600		
450			610		
455			620		
460			630		
465			640		
470			650		
475			660		
480			670		
490			680		
500			690		
510			700		
520					

Estimated amount of chlorophyll a/chlorophyll b _____

Attach spectrum and indicate maximum wavelengths.

Instructor's Signature _____

Data and Results (Chlorophyll)

Part C: Paper Chromatography of Ethanol Extract

Record your observations using tables and/or diagrams. You should also attach filter paper with spots.

Questions

1. Why is it necessary to shred the spinach before extraction?

2. How do plants benefit from the presence of both chlorophyll a and b?

3. Why would you expect to find very little β-carotene in your spinach extract?

4. What combination of household products could be used as a substitute for petroleum ether mixed with acetone in order to separate chlorophyll a and b using paper chromatography?

Experiment 8
Measuring pH with Universal Indicator

Name_____ Date_____ Section_____

1. Vinegar is a 5% solution of acetic acid. Which is lower, the pH of vinegar or the pH of a 5% muriatic acid, HCl (aq), solution? Answer without doing any calculations. Give your reasoning.

2. A solution is prepared by adding 0.40 g lye to water to make 1 L solution. Assuming that the lye is 100% sodium hydroxide, find the:
a) molarity   b) pOH   c) pH

3. The concentration of a typical sample of household ammonia is 51 g/L solution. Find the molarity of the solution. Calculate its pH given the value of 1.6×10^{-5} for $K_b NH_3$.

Instructor's Signature _____

Experiment 8
Measuring pH with Universal Indicator

PURPOSE
Anthocyanin will be removed from red cabbage by heating with alcohol. The concentrated extract will be used to measure the pH of various solutions.

INTRODUCTION
How acid-base indicators work is given. The remainder of the discussion is about universal indicators including a detailed description of anthocyanin.

Acid-Base Indicators
An acid-base indicator can be used to make an approximate measurement of pH. In the aqueous ionization below, the acidic indicator, HIn, donates a proton to water to produce the anion, In^-, and hydronium, H_3O^+:

$$HIn + H_2O \rightleftharpoons In^- + H_3O^+$$

For an indicator to be effective, the unionized and ionized forms must be different colors (or one form can be colorless and the other colored). A given Indicator changes color over a specific range of pH values. For example, phenolphthalein dye changes from colorless to pink in the range from 8 to 10. The dye known as litmus turns red in acids and blue in bases.

Universal indicators (dyes or mixtures of dyes) change color every pH unit or so, making them particularly convenient for measuring pH. One commercially available product responds between pH 4 and 10. A single indicator can thus be used to measure the pH of a broad range of solutions.

Anthocyanin
Anthocyanin, present in purplish fruits, vegetables and flowers, is a natural universal indicator. The formula below is the core structure of anthocyanin. Different plants have different sugars attached to the OH groups labeled 3, 5, and 7.

pH	Color
1-3	red
4	pink
5-6	violet-pink
7	violet
8	blue
9	blue-green
10-12	green-blue

Anthocyanin from red cabbage gives the color changes listed which are particularly clear and easy to see.

HAc/NaAc Buffer

A buffer solution (resists change in pH) can be made from the weak acid, acetic acid, combined with its salt, sodium acetate. The pH of an equimolar HAc/NaAc buffer is the same as pK_a for the acidic component, HAc. The value of K_a at 25°C is 1.75×10^{-5}.

$$pK_a = -\log(1.75 \times 10^{-5}) = 4.8$$

APPARATUS

A hotplate and assorted glassware is sufficient.

SAFETY

Wear safety glasses throughout this experiment.

Chemical	Toxicity	Flammability	Exposure
70% isopropyl alcohol	1	2	0
0.01 M HCl	1	0	1
0.01 M CH_3COOH	0	0	0
0.01 M NaOH	1	0	1
0.01 M NH_3	2	0	2
5% acetic acid (vinegar)	0	0	0
washing soda	1	0	1
baking soda	1	0	1
borax	0	0	0
NaAc/HAc buffer	0	0	1

0 is low hazard, 3 is high hazard

PROCEDURE

Anthocyanin indicator is extracted from red cabbage with alcohol (Part A). If the indicator is made ahead for you, start with Part B where the concentrated extract is used to measure the pH of various products. A buffer solution is prepared and tested using anthocyanin as the indicator in Part C.

Part A: Extraction of Anthocyanin from Red Cabbage

1. Mix 150 g very finely chopped red cabbage and 300 mL rubbing alcohol in a 600-mL beaker and boil 1 hour.

2. Filter and reduce the 300 mL liquid to 30 mL of deep purplish-black colored extract.

Part B: Measuring pH

1. Strong vs. Weak Acids (and Bases)
Place 5 -10 mL 0.01 M HCl in a small beaker.
Add 5-10 drops of indicator. Record the color and corresponding pH. Repeat for 0.01 M HAc, 0.01 M NaOH and 0.01 M NH_3.

2. Generic vs. Name Brand Products
Compare the pH of the two vinegar and washing soda samples.

3. Measure pH of assorted products including baking soda and borax. Record color changes and any other observation.

Part C: Preparation and Testing of NaAc/HAc Buffer
You will prepare a buffer solution, then compare its behavior to that of plain water. Dilute acids and bases will be available for the testing procedure.

1. Prepare sodium acetate/acetic acid, NaAc/HAc, buffer by adding 1.0 g NaOH to 50 mL vinegar (HAc). This makes a buffer solution that is approximately equimolar.

2. Create a chart listing six tests that can be used to compare the buffer with plain water before and after the addition of small amounts of dilute acid and base. Use anthocyanin as the indicator. Record and explain color changes.

Data and Results (pH)

Name_____ Date_____ Section_____

Part B: Measuring pH

Sample	Indicator Color	pH

Observations:

Instructor's Signature _____

Data and Results (pH)

Part C: Preparation and Testing of NaAc/HAc Buffer

Prepare a chart to record your data.

Questions

1. To the nearest 0.5 unit, calculate the pH of 0.01 M HCl and 0.01 M HAc acid (K_a 1.8 x 10^{-5}). Compare to your experimental values.

2. To the nearest 0.5, calculate the pH of 0.01 M NaOH and 0.01 M NH$_3$ (K_b 1.6 x 10^{-5}). Compare to your experimental values.

3. Could anthocyanin distinguish 0.01 M HCl from 0.02 M HCl? Explain.

4. Compare the pH of brand name products to the corresponding generic ones.

5. Show that the buffer solution prepared in Part C is *approximately* equimolar, assuming that the concentration of acetic acid in vinegar is roughly 1 M.

Experiment 9
Extraction of Curcumin from Turmeric

Name _John Gilligan_ Date _3/14/15_ Section _61_

1. Curcumin, present in the spice turmeric, is soluble in only one of the following solvents. Which is it? Explain.

a) CH₃CH₂OH (ethanol)
b) C₆H₁₄ (hexane)
c) H₂O

Curcumin is soluble in ethanol because it can form Hydrogen bonds with ethanol (they both have alcohol functional groups). Curcumin can not dissolve in water because even though it is polar overall, some areas of Curcumin are nonpolar and will not dissolve in water (see O)

2. The absorbance of a solution at 500 nm is 0.40. If the molar absorptivity, ε, is 5.0 cm⁻¹ mol⁻¹L and the path length is 1 cm, use Beer's Law to find the concentration of the solution. At what concentration would deviation from Beer's Law be expected?

$A = E \cdot b \cdot C$
$0.4 = 5cm^{-1}mol^{-1}L \cdot 1cm \cdot C$

$c = 0.08 M$

Beer's law does not work when $A \geq 1$ since it does not represent a linear equation.
So: $1 \leq 5cm^{-1}mol^{-1} \cdot 1cm \cdot C$
$c \geq 0.2M$

3. To be used as a pH indicator, what property must a compound have?

A compound must go through a physical change (e.g. color change) when there is a notable pH change (e.g. when an acid becomes a base) in order to be used as a pH indicator.

Instructor's Signature _____

Experiment 10
pH-Titration

Name_____ Date_____ Section_____

1. A 25.0 mL sample of unknown hydrochloric acid solution is titrated with 0.100 M NaOH. The equivalence point was reached after the addition of 23.5 mL NaOH.
a) Is the concentration of acid greater or less than 0.100 M.? (No calculations should be needed for this part)
b) Find the concentration of the unknown HCl(aq).

2. Find pK for the following ionization constants, K =
a) 1.8×10^{-5} b) 1.1×10^{-2}

3. For the ionization of a weak acid, HA:
a) Give the expression for K_a.
b) After taking the $\log_{10}$ of both sides of the equation above, solve for pH. (This is known as the Henderson-Hasselbalch equation).
c) Under what conditions would pH be equal to pK?

Instructor's Signature _____

Experiment 10
pH-Titration

PURPOSE
A pH-meter will be used to follow the titration of acids with sodium hydroxide, NaOH(aq).

INTRODUCTION
Unknown acids are titrated with a standard NaOH(aq) solution with known concentration

Standard NaOH(aq)
The standard NaOH(aq) is prepared with commercial lye sold as oven and drain cleaner. As long as the container is used within a few weeks of being opened, the quality of NaOH is similar to that of the laboratory reagent.

Strong Acids
When HCl(aq) is titrated with NaOH(aq) the neutralization reaction is:

$$HCl(aq) \ + \ NaOH(aq) \ \rightarrow \ NaCl(aq) \ + \ H_2O$$

The titration is performed by adding small amounts of standard NaOH to HCl of unknown concentration. The pH is recorded then plotted vs. the volume of NaOH added to the HCl solution. The result of this plot is an "S" shaped curve. The middle of the "S" is the point when moles of acid are equivalent to moles of base--- at pH 7. From the equivalence point, the concentration of the unknown acid, HCl, can be found.

Weak Acids
Weak acids, HA, are titrated with NaOH solution using the same technique.

$$HA(aq) \ + \ NaOH(aq) \rightarrow \ NaA(aq) \ + \ H_2O$$

However, the equivalence point of a weak acid vs. strong base titration is at an alkaline pH. The concentration of the unknown acid is once again found using the equivalence point taken from the middle of the "S" curve when moles of acid are equivalent to moles of base.

The equilibrium constant for weak acids, K_a can be determined from the titration data. The pK_a (-log K_a) corresponds to the pH at *half* the equivalence point volume. To see why this is true, examine the expression for the ionization of HA:

$$HA \ + \ H_2O \ \rightleftharpoons \ A^- \ + \ H_3O^+$$

$$K_a = \frac{[A^-][H_3O^+]}{[HA]}$$

When [HA] = [A⁻], K_a = H_3O^+, - log pK_a = -log H_3O^+, and so, pK = pH.   At the half equivalence point half the HA has undergone neutralization producing an equal amount of A⁻ ion. Thus, at the half equivalence point [HA] = [A⁻] and pK_a = pH.

Indicator
For visual evidence of the equivalence point, an indicator is added to the acid sample. The pH when the color change occurs is the "end point". The range of the indicator may not even include the equivalence point. For example, phenolphthalein changes from colorless to pink in the range pH 8 to 10, which does not include 7. However, since the pH increases abruptly by as much as 4 units upon the addition of a fraction of a drop at the equivalence point for titrations of strong acids, it makes no difference.

Curcumin is another indicator that changes color in the range from 8 – 9 (yellow to red). See Experiment Number 9. Since phenolphthalein is no longer available in over the counter products, curcumin will be used in this experiment.

Polyprotic acids
Polyprotic acids have more than one proton that ionizes. For instance, a weak diprotic acid, H_2A, ionizes in two steps

$$H_2A + H_2O \rightleftharpoons HA^- + H_3O^+ \qquad K_{a1}$$

$$HA^- + H_2O \rightleftharpoons A^{2-} + H_3O^+ \qquad K_{a2}$$

The value of K is smaller for each successive ionization and so K_{a1} is greater than K_{a2}.

An inflection appears in the titration curve for each ionization, as long as the corresponding values of K are separated by at least 10^{-3}, that is, 3 pK units, and provided the ionization is not too weak. For instance, equivalence points for the first two ionizations for phosphoric acid can be seen, but not the third (See Table 10.1).

APPARATUS

The titration apparatus includes a pH meter, hydrogen electrode and a buret containing the standard base. The unknown acid sample is in the beaker on the stir plate.

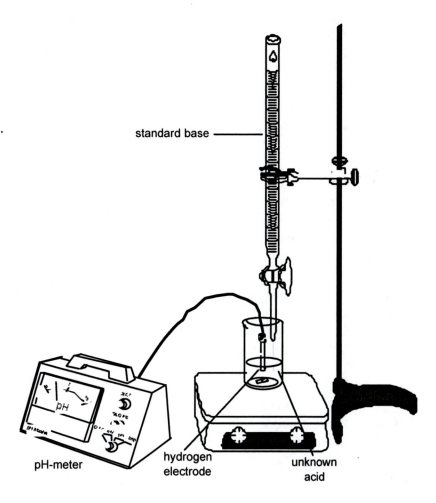

standard base

pH-meter

hydrogen electrode

unknown acid

SAFETY

Wear gloves throughout this experiment. Curcumin is a yellow dye that stains the skin and is slightly hazardous in case of skin or eye contact (irritant).

Chemical	Toxicity	Flammability	Exposure
curcumin	1	0	1
NaAc/HAc buffer	0	0	1
0.1 M NaOH	1	0	1
0.5 M or less HCl	1	0	1
phosphoric acid (< 1%)	1	0	1
citric acid	0	1	0
ascorbic acid	0	0	0
acetic acid (< 3%)	0	0	0

0 is low hazard, 3 is high hazard

PROCEDURE
Part A: Calibration of the pH-Meter with Buffers
The pH meter is calibrated by using one or more buffer solutions of known pH. Instructions below are for analog meters. The buffers used here are equimolar NaAc/HAc, pH = 4.76 and 0.10 M Borax, pH = 9.2.

1. Turn the OFF/pH/mV knob to the OFF position. Plug in the meter. Mount the electrode and connect the jack to the pH meter. Set the temperature dial on the back of the pH meter to the actual temperature in the lab

⚠**CAUTION:** The plastic electrode body *must* be kept moist at all times or it will be ruined. Do not touch the bulb with anything.

2. Place 7 to 10 mL of the buffer solution in a 6-inch test tube. Label it with the pH value.

3. Immerse the electrode into the buffer solution. In order to obtain accurate readings, the electrode has to be immersed at least up to the little electrochemical junction on its side (i.e. about 1.5 inch into the solution), but no further than about two-thirds of the way up the electrode body (otherwise you may short-circuit it).

4. Set the OFF/mV/pH knob to pH.

5. Use the SET knob to dial the pH of the buffer (i.e. set it to 4.25). *From this point on, do not change this setting since changing it would alter your calibration!* (Save the buffer in case you wish to recheck the calibration.)   If the pH meter is not in use turn the OFF/mV/pH knob to the OFF position. If you need to unplug the electrode jack, turn the meter OFF first !

Part B: Titration of Unknown HCl with Standard (Known) NaOH

1. Place about 200 mL of standard NaOH solution in a clean, 250-mL Erlenmeyer flask. Record the molarity of the NaOH(aq).

2. Clean a 50-mL buret thoroughly with tap water, then rinse it with several small portions of the standard NaOH solution, being sure to run some solution through the tip.

3. Fill the buret above the '0' mL mark, then lower the meniscus back to '0'.

4. Using a volumetric pipet and a pipet pump (or bulb) measure 25.00 mL of unknown HCl solution into a 150-mL beaker. Place the beaker on a stir plate and clamp the electrode so that it is supported in the solution.

5. Place a magnetic stir bar in the beaker so that it does not touch the electrode while stirring.

6. Add 10 drops of curcumin indicator to the HCl solution.

7. Place the buret just inside the beaker as shown in the figure so that the buret will drip directly into the solution.

8. Trial run: Add NaOH in roughly 2 mL volumes until the endpoint is reached, that is, when the curcumin changes from yellow (acidic) to red (basic). Record the approximate volume of NaOH used to reach this point. The purpose of the trial run is to find the approximate end point so you will know when to slow down during your careful run (next step). The trial run should take only a minute or two.

9. Careful run: Add NaOH gradually (1 mL at a time) until you approach within 1 mL of the estimated endpoint. Then proceed much more slowly drop by drop. After the endpoint is reached, measure 5-10 more points adding 1 mL of NaOH at a time, to see the titration curve leveling off again. With practice you should be able to deliver half drops near the end point.

10. Perform another careful run with a second sample of the same unknown HCl. If time allows, you could try a third careful run.

11. Plot the pH vs. the volume of NaOH added. Find the concentration of the unknown acid. The equivalence point is at the inflection point of the S curve, where the pH change is the steepest.

Part C: Titration of Unknown Weak Acid (HA) with Standard (Known) NaOH

1. Repeat the pH-meter calibration.

2. Repeat Part A using 25.00 mL of the unknown weak acid (instead of HCl). The unknown acid will have a concentration of approximately 0.1 M. The equivalence point of a weak acid/strong base titration is at an alkaline pH. Curcumin, which changes color within the range 8-9, can be used as the indicator.

3. Plot pH vs. volume of NaOH added. Find the concentration of the unknown acid and values for pK_a and K_a. Recall that the pK_a corresponds to the pH at half the equivalence point volume. Be sure to gather some pH data around this volume.

Part D: Acids in Consumer Products

Titrate one of the products available using standard NaOH. Use the shape of the titration curve to help identify the acid it contains. Recall that two K values must be far enough apart (at least 10^{-3}) to see two equivalence points and that very weak ionizations will not produce an equivalence point. Record data and titration curve.

Table 10.1 Values of K at 25°C for some diprotic and triprotic acids. Citric and phosphoric acids are found in soft drinks.

Acid	Formula	K_{a1}	K_{a2}	K_{a3}
Phosphoric	H_3PO_4	7.8×10^{-3}	6.2×10^{-8}	4.8×10^{-13}
Citric*	$C_6H_8O_7$	7.4×10^{-4}	1.7×10^{-5}	4.0×10^{-7}
Oxalic*	$C_2H_2O_4$	6.5×10^{-2}	6.1×10^{-5}	

*Formulas are shown below with acidic H's in bold.

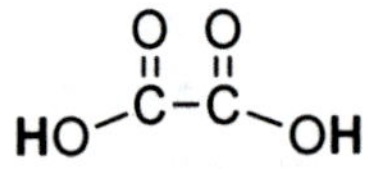

oxalic acid 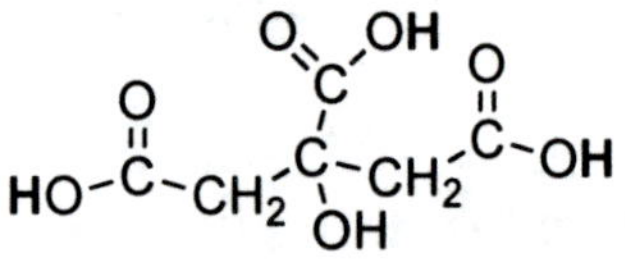   citric acid

Data and Results (pH-Titration)

Name_____ Date_____ Section_____

Part B: Titration of Unknown HCl with Standard (Known) NaOH

Molarity of standard NaOH _____ M

Code number/letter of unknown HCl _____

Trial run: Approximate equivalence point _____ mL NaOH

Careful Runs

Titration 1				Titration 2			
pH	NaOH(mL)	pH	NaOH(mL)	pH	NaOH(mL)	pH	NaOH(mL)

Equivalence point 1 _____ mL NaOH

Equivalence point 2 _____ mL NaOH

Average of 1 and 2 _____ mL NaOH

Concentration of HCl _____ M

Instructor's Signature _____

Data and Results (pH-Titration)

Part C: Titration of Unknown Weak Acid (HA) with Standard NaOH

Molarity of NaOH _____ M

Code number/letter of unknown weak acid _____

Trial run: approximate equivalence point _____ mL NaOH

Titration 1				Titration 2			
pH	NaOH(mL)	pH	NaOH(mL)	pH	NaOH(mL)	pH	NaOH(mL)

Equivalence point 1 _____ mL NaOH

Equivalence point 2 _____ mL NaOH

Average of 1 and 2 _____ mL NaOH

 Concentration of HA _____ M

pK_a (pH at half-equivalence volume) _____

K_a _____

Part D: Acids in Consumer Products

Questions

1. Why can curcumin be used as the indicator for titrations of NaOH vs. either strong or weak acids?

2. Why should the volume of standard base (or acid) needed to reach the equivalence point in a titration be more than 10.0 mL?

3. The titration curve shown is for 100 mL 0.10 M weak acid titrated with 0.10 M NaOH. Find:
a) the volume of NaOH required to reach the equivalence point
b) pK_a (to the nearest whole number)
c) K_a

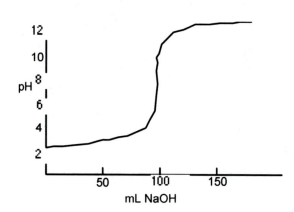

Experiment 11
Preparation & Acid-Base Properties of Ammonium Chloride

Name_____ Date_____ Section_____

1. Write the ionization reactions for aqueous solutions of:
a) ammonia   b) ammonium

2. Give a balanced equation for the neutralization of hydrochloric acid with sodium hydroxide. Suppose 1.0 mol sodium hydroxide reacts with an excess of hydrochloric acid to produce 55 g sodium chloride. What is the yield of NaCl?

3. Find the molarity of 20% muriatic acid, HCl(aq), assuming that the density of the solution is 1.0 g/mL.

Instructor's Signature _____

Experiment 11
Preparation & Acid-Base Properties of Ammonium Chloride

PURPOSE
Aqueous ammonia will be mixed with hydrochloric acid to produce ammonium chloride using amounts determined via titration of reactants. The reaction yield will be found and the acid-base properties of aqueous ammonium chloride studied.

INTRODUCTION
The synthesis of ammonium chloride is a common experiment in chemistry classes in Finland, where residents are fond of salmiakki, a salty treat that resembles licorice in color. Its main ingredients are ammonium chloride, licorice root and usually black coloring. Only candies in which NH_4Cl is a main ingredient can be called salmiakki, derived from salmiac, a common name for ammonium chloride. Extremely popular in all five of the Nordic countries as well as the Netherlands, salmiakki is uncommon outside Europe where it is often intensely disliked by most who taste it for the first time. Salmiakki is also used to flavor liquors and dipping sauces.

Synthesis of Ammonium Chloride
Ammonia and hydrochloric acid react to produce ammonium chloride:

$$NH_3(aq) \ + \ HCl(aq) \ \longrightarrow \ NH_4Cl(aq)$$

Sources for the reactants are household cleaner, $NH_3(aq)$ and muriatic acid, 20% HCl.

The amounts to combine can be found by titrating a given volume of ammonia cleaner with the muriatic acid and noting the end point using an indicator such as anthocyanin. Depending on the amount of product desired, the corresponding volumes (or fractions of the volumes) are then combined to run the reaction, this time, without the indicator.

Acidity of NH₄Cl(aq)
Aqueous ammonium chloride is acidic. The ammonium ion donates a proton to water according to the ionization:

$$NH_4^+ \ + \ H_2O \ \rightleftharpoons \ NH_3 \ + \ H_3O^+$$

The expression for K_a, 5.6×10^{-10} at 25°C, is: $\quad K_a = \dfrac{[NH_3][H_3O^+]}{[NH_4^+]}$

Ammonium chloride is used in electroplating, tinning and the manufacturing of dyes. Solutions of NH_4Cl are also used to treat severe cases of metabolic alkalosis, caused possibly by ingestion of lye (NaOH) or some other base.

NH_4Cl/NH_3 Buffer

A buffer solution (resists change in pH) can be made from the weak base, ammonia, combined with its salt, ammonium chloride. The pH of an equimolar NH_3/NH_4Cl buffer is the same as pK_a for the acidic component, NH_4^+. The value of K_a at 25°C is 5.6×10^{-10}.

$$pK_a = -\log(5.6 \times 10^{-10}) = 9.2$$

APPARATUS

A titration will be used to find the right amounts of reactants to use in the synthesis of ammonium chloride. The hydrochloric acid is in the buret and the ammonia solution is in the Erlenmeyer flask.

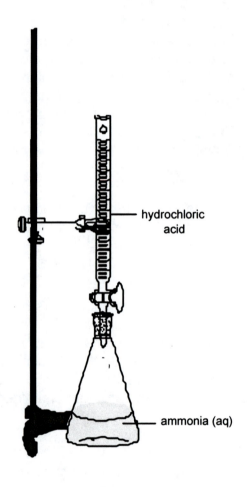

hydrochloric acid

ammonia (aq)

SAFETY
Wear safety glasses throughout this experiment.

Chemical	Toxicity	Flammability	Exposure
muriatic acid	2	0	2
1.5 M ammonia	2	0	1
anthocyanin indicator	0	0	0
ammonium chloride	0	0	1
ammonium chloride solution	0	0	1
NH_3/NH_4Cl buffer	1	0	2
0.1 M NaOH	1	0	1
0.1 M HCl	1	0	1

0 is low hazard, 3 is high hazard

PROCEDURE
In Part A, the titration of ammonia and HCl determines the correct volumes of solutions to combine. The reaction is run in Part B.   In Part C students study the acid-base properties of the product, ammonium chloride.

Part A: Finding Amounts of Reactants

1. Fill the 50-mL buret with muriatic acid above the '0' mL mark, then lower the meniscus back to '0'.

2. Measure 50 mL of household ammonia solution into a 250-mL Erlenmeyer flask.

3. Add 5 drops of anthocyanin indicator to the ammonia solution.

4. Place the buret just inside the flask as shown in the figure so that the buret will discharge directly into the solution.

5. Add acid while swirling the flask until the end point is reached, that is, when the indicator changes from yellow to green to blue to dark pink. Record the volume of muriatic acid used to reach this point.
Recall that the equivalence point for a strong acid vs. weak base titration occurs at pH < 7.

Part B: Reacting Ammonia with Hydrochloric Acid

1. Put 10 mL of the household ammonia solution in a 50-mL beaker. This is 1/5th the volume of the sample ammonia volume used in the titration.

2. Add 1/5th the volume (this will probably be about 2-3 mL) of muriatic acid that was needed to change the color of the indicator. Bring the mixture to a boil and allow liquid to evaporate until about half the volume is left.

3. Pour the mixture into an evaporating dish and heat on a hotplate at a low setting until the water evaporates.

⚠ **CAUTION:** Make sure the hotplate is at a low setting. Overheating can cause the evaporating dish to crack. If you notice "spitting", reduce the heat even lower.

4. Scrape the white solid onto a weighing paper making sure to get all of it. Allow to dry in air. Weigh several times until the mass remains the same.

5. Find the theoretical mass of ammonium chloride expected assuming that the concentration of the muriatic acid was about 5.5 M. From the actual mass of NH_4Cl produced, find the approximate %yield.

Part C: Acid-Base Properties of NH₄Cl Solution

1. Using your dry product, prepare 25 mL 0.15 M NH_4Cl solution and measure the pH using pH paper.

2. Prepare a buffer solution by mixing equal volumes of 0.15 M ammonia solution and 0.15 M NH_4Cl. Measure the pH. The household ammonia is about 1.5 M.

3. Design a procedure to test the ability of the buffer solution to resist pH changes. Describe your method and record the results. Include tables and/or plots as needed.

Data and Results (Ammonium Chloride)

Name_____ Date_____ Section_____

Part A: Finding Amounts of Reactants

Volume household ammonia __50__ mL

Volume muriatic acid needed to reach the end point _____ mL

Part B: Reacting Ammonia with Hydrochloric Acid

Vol HCl to react _____ mL

Vol ammonia to react _____ mL

Molar mass NH_4Cl _____ g/mol

Theoretical mass NH_4Cl expected _____ g

Actual mass obtained _____ g

Yield _____ %

Part C: Acid-Base Properties of NH_4Cl Solution

Instructor's Signature _____

Questions

1. Could curcumin or phenolphthalein be used as an indicator for the titration of $NH_3(aq)$ with $HCl(aq)$? Explain.

2. Explain how the amounts of reactants to use in the synthesis of ammonium chloride were determined.

3. The amount of ammonium chloride obtained will be less than the maximum theoretical amount, that is, the yield will be less than 100%. What are some reasons for this?

4. Suppose you need to make at least one gram of ammonium chloride. What volume of each starting material would you use?

Experiment 12
Boyle's Law

Name_____ Date_____ Section_____

1. Convert the following units of pressure to mmHg:
a) 3.00 atm   b) 1060 cm H_2O c) 101 kPa

2. The quantity pressure is the ratio of force to the surface area over which it is exerted. Show that force/area is not a unit of length. Why are pressures often expressed in length units?

3. Suppose there is only one way to measure the mass of air contained in a 1-L flask, that is, by using a top-loading electronic balance. What difficulties would be encountered?

Instructor's Signature _____

Experiment 12
Boyle's Law

PURPOSE
The relationship between pressure and volume at constant temperature, Boyle's Law, will be found by collecting data on a sample of air.

INTRODUCTION
It was the invention of the mercury barometer by Evangelisto Torricelli (1608-1647) In 1643 that would lead to the discovery of the pressure volume relationship of gases. Since mercury is more than 13 times denser than water, the height of a column of mercury that would be supported by sea level pressure is 76 cm compared to 34 ft for a column of water. Using a mercury barometer makes it practical and convenient to measure the volume of a gas over a wide range of pressures.

Almost 20 years after the invention of the mercury barometer, Robert Boyle (1627-1691) published the law relating the pressure of a gas to its volume at constant temperature in 1662. Boyle would have gone on to study the volume vs. temperature relationship, but at the time there was no thermometer and no temperature scale available. More than a century would go by before a law relating the volume and temperature of a gas was published.

The relationship Boyle found between the pressure and volume will be reproduced in this experiment. The pressure of the air to be studied is the atmospheric room pressure plus the height of liquid in a barometer tube (expressed in the same units). Although it is true that the water barometer used in this experiment severely restricts the range of pressures that can be measured, the data collected will still be sufficient to illustrate clearly the relationship between pressure and volume.

APPARATUS
In Part A a barometer is used to measure air pressure using the setup shown along with a meter stick to find the height of liquid. If possible, the liquid in devices such as the set-up shown here should have a very low vapor pressure at room temperature. Of the best candidates, mercury itself (v.p. = 0.002 mmHg) is toxic, di-n-butyl phthalate (v.p. = 0.0001 mmHg) is not readily available and mineral oil (v.p. < 0.01 mmHg) is easily found, but responds too slowly. Therefore, water will be used. Its vapor pressure (20 mmHg) will not interfere at all with the determination of the pressure volume relationship for air.

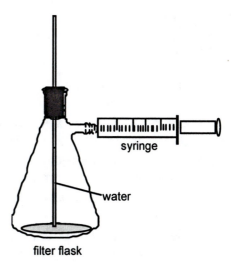

syringe

water

filter flask

In Part B air pressure is measured using a gas pressure sensor probe. The gas pressure sensor probe will be connected to a Vernier interface panel, which is connected to a laptop computer. The syringe is attached directly to the pressure sensor.

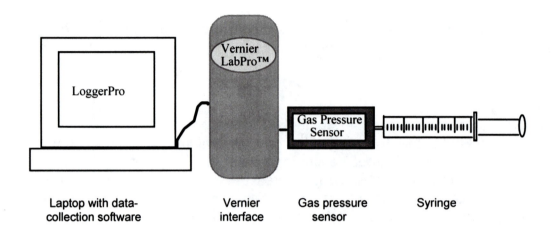

| Laptop with data-collection software | Vernier interface | Gas pressure sensor | Syringe |

SAFETY
Wear safety glasses throughout this experiment.

Chemical	Toxicity	Flammability	Exposure
water with food coloring	0	0	0

0 is low hazard, 3 is high hazard

PROCEDURE
Record the temperature and barometric pressure.
Part A uses the filter flask apparatus with syringe. Handle the flask as little as possible during the procedure to avoid changing the temperature of the system.
In Part B a computer-interfaced method with a gas sensitive pressure probe/computer and syringe is used.

Part A: Measuring Volumes and Corresponding Pressures (Barometer)

1. Measure 25.0 mL of colored (dyed) water with a 25-mL volumetric pipet or a graduated cylinder into a 500-mL filter flask. Close the system by replacing the top of the apparatus. The syringe plunger must be all the way in. The bottom of the tube should be approximately 1 mm under the surface of the water when the rubber stopper is airtight. Make sure your apparatus has no leaks.

⚠**CAUTION:** Do not attempt to move the glass tube up or down through the rubber stopper. It can break in your hand causing serious cuts.

2. Before recording data, the ruler must be placed close enough to the apparatus to prevent any height recording error. When starting the experiment, the syringe should be pulled back to the 30 mL (30 cc) mark. This will also remove any bubbles in the fluid column. Wait for about five minutes to allow the system to come to equilibrium.

3. Obtain PV data by pressing the syringe plunger to a new position in increments of 5.0 mL. The height of the water column is measured from its surface and not from the tabletop. Allow one minute for your system to come to equilibrium after each reading. If the water column is hidden by the stopper you can either change the syringe volume a little or skip that point altogether. Make sure the syringe does not move during a reading; this tends to be a problem when the pressure increases.

4. The "uncorrected" flask volume, $V_{(unc)}$, includes the 500-mL filter flask with syringe completely plunged in and the top on. Fill the flask completely with water and replace the top of the apparatus allowing excess water to flow out. Once the top is on and there are no bubbles present in the flask, the volume can be determined. Measure the water in the flask by transferring it into a graduated cylinder. Record as $V_{(unc)}$.

5. Repeat for the 2nd trial.

6. Treatment of Data:
  a. Volume of air in the plastic tube, $V_{(p.tube)}$, connecting syringe with flask:

$$V_{(p.tube)} = \pi \times r^2 \times l$$

where r = inner radius and l = length of plastic tube section filled with air.

  b. Volume of air in the flask, V flask. For $V_{(unc)}$ see Step 4 above.

$$V_{(flask)} = V_{(unc)} + V_{(p.tube)} - 25.0 \text{ mL}$$

  c. Volume of air, V:

$$V = V_{(flask)} + V_{(syr)}$$

  d. Pressure of the air, P (in cmHg)

P = Barometric Pressure (cm Hg) + Height of water (cm)/13.56

  e. Plot P vs. V and P vs. 1/V for both trials.

Be sure the recorded data includes only what is measurable. For example, the pressure can be read only to the nearest mmH$_2$O.

Part B: Using a Gas Pressure Sensor Interfaced with a Computer

1. Plug in the Vernier interface to an outlet. You will hear a beep. Connect the gas pressure sensor probe to the channel 1 port of the Vernier interface. Attach the interface to the back of the computer using the usb cord.

2. Open the LoggerPro software application on the computer desktop. LoggerPro will detect the gas pressure sensor probe. An experiment file for collecting gas pressure data will appear.   Now you are ready to take gas pressure readings. Select 0 to 1600 mmHg for the pressure range using the Setup pull-down menu.

⚠**CAUTION:** The maximum pressure that the sensor can tolerate without permanent damage is about 3 atm.

3. Set the plunger of the syringe to the 10-mL mark, using the top of the black rubber fitting to mark the volume.

4. Connect the plastic syringe to the white stem on the end of the Gas Pressure Sensor. The white stem has a small threaded end called a luer lock.   Connect the plastic syringe to the white stem *gently*.

⚠**CAUTION:** It is very easy to damage the connector.

5. Click on the Collect tab. This will open a menu with the options Stop and Keep. Press the Keep tab. When the window opens prompting you to Enter a Number, enter the volume to which the syringe is set and press OK. The Pressure-Latest reading from the Meter Window will be plotted in the Graph Window at that volume.
One member of the group can adjust the syringe volume while another enters the value.

6. Repeat step 4 for every 1 mL increment between 4 and 20.

⚠**CAUTION:** Do not try to decrease the volume below 4 mL. The maximum pressure that can be recorded without damaging the sensor is about 3 atm.

7. When complete click on Stop and Print Screen.
Record PV data on the data sheet.   Plot P vs. V   and P vs. 1/V.
Be sure the plotted points reflect the uncertainty in the measurement. For example, the pressure sensor reads to the nearest 0.40 mmHg. The volume reading is to the nearest 0.2 mL.

Data and Results (Boyle's Law)

Name_____ Date_____ Section_____

Part A: Measuring Volumes and Corresponding Pressures (Barometer)

Internal radius of plastic tube, r (estimate with ruler): _____ cm

Length of plastic tube filled with air, l _____cm, V(p.tube) _____mL

V(unc) _____ mL   V(flask) _____mL

Barometric Pressure _____ mm Hg   Temperature _____ °C

1st trial

V Syringe (mL)	V Air (mL)	V Air (L)	Height of Liquid (cm H$_2$O)	Height of Liquid (cm Hg)	Pressure Air, P (cm Hg)	P V (cm Hg x L)

2nd trial

V Syringe (mL)	V Air (mL)	V Air (L)	Height of Liquid (cm H$_2$O)	Height of Liquid (cm Hg)	Pressure Air, P (cm Hg)	P V (cm Hg x L)

Instructor's Signature _____

Data and Results (Boyle's Law)

Part B: Using a Gas Pressure Sensor Interfaced with a Computer

Barometric Pressure _____ mm Hg

Temperature _____ ˚C

V Air (mL)	V Air (L)	Pressure (mmHg)	P V (mm Hg x L)

Attach printout of P vs. V and your plot of P vs. 1/V.

Questions

1. The procedure suggests taking volume measurements in 5-mL (or 1-mL) increments. Would it matter if one (or more) of the volumes chosen were not 5 mL (or 1 mL) from the previous one? Explain.

2. What volume mark on the syringe would correspond to a pressure of 3.0 atm? (You can use a set of data points and Boyle's Law, $P_1V_1 = P_2V_2$)

3. Under what conditions would you expect the PV product to vary most from the average PV? Why?

4. Boyle's law combined with Charles' law relating temperature and volume and Avogadro's law stating that volume is proportional to number of molecules (and moles) led to another relationship, PV = nRT. This is the *ideal gas law* in which n is number of moles and R is the *gas constant* (0.0821 L atm/mol K). How could you use the ideal gas law to find the mass of air in a 1-L flask at room temperature and sea level? (Compare to method in Prelab Question #3)

Experiment 13
Molar Mass of a Volatile Liquid by Vapor Density

Name_____ Date_____ Section_____

1. Assuming ideal behavior, what is the approximate volume in liters at room temperature and pressure of one mole of:
a) nitrogen gas   b) water vapor

2. Starting with the ideal gas law, derive an expression for the molar mass, M.

3. Starting with the ideal gas law, derive an expression for the density of a gas.

Instructor's Signature _____

Experiment 13
Molar Mass of a Volatile Liquid by Vapor Density

PURPOSE
The molar mass of a volatile liquid will be found by measuring the mass of vapor produced upon heating a liquid in a container of known volume, at a known temperature and atmospheric pressure. The ideal gas law is used to make the calculation.

INTRODUCTION
French chemist, Jean-Baptiste-André Dumas (1880-1884), developed methods for analysis of chemical compounds including the use of vapor density for finding the formula of a volatile liquid.

Dumas Method
The mass of a known volume of vapor is found by weighing a flask before the liquid is added and once again after the flask is filled with vapor from heating the liquid. The molar mass (M) is calculated by applying the vapor density (m/V) form of the ideal gas law:

$$PV = nRT$$

$$PV = \frac{m}{MRT}$$

$$M = \frac{mRT}{PV}$$

where:
m = mass of liquid that vaporizes to fill flask
T = temperature of water bath
P = barometric pressure
V = volume of flask

When the liquid in a flask is heated in a water bath the liquid vaporizes through a tiny hole until the flask is filled with vapor. Using the data above and a correction for lost air discussed next, the molar mass (M) of the liquid can be calculated.

Correction for "Lost Air"
In this experiment the mass of vapor will be found by subtracting the mass of the original "empty" flask from that of the final flask filled with the liquid that vaporized. The mass of the original vessel included *all* the air it contained. When the liquid is added to the flask the vapor it produces will force some of that air out of the flask through the hole. This happens because the internal pressure in the flask must remain equal to the pressure in the room. That means that the amount of air in the final flask will be *less than it was in the original flask.* This "lost air" was part of the measured mass of the original flask and therefore must be added to the mass of the final flask. Otherwise the mass of the volatile liquid will be too low, and so will the calculated molar mass.

The number of moles of vapor produced is equal to the number of moles of air driven off. The ideal gas law can be used to calculate moles of vapor where T is the room temperature, V is the volume of the flask and P is the vapor pressure of the liquid at room temperature.

$$PV = nRT$$

$$n_{vapor} = \frac{P_{vapor}V}{RT} = n_{air}$$

This gives the number of moles of air driven off, which can be converted to mass using the average molar mass for air, 29 g/mol. The number of grams of lost air is then added to the mass of the final flask.

The table below includes vapor pressures of common liquids at various temperatures.

Table 13.1 Vapor pressures (mmHg) of common liquids from 18 to 25°C*

Liquid	18°C	19°C	20°C	21°C	22°C	23°C	24°C	25°C
acetone	148	154	162	169	177	185	193	201
methanol	80	84	89	93	98	104	109	115
ethanol	37	39	41	44	46	49	52	55

*Estimate vapor pressures at temperatures not listed.

APPARATUS
As the liquid in the flask is heated in a water bath it vaporizes through a tiny hole until the flask is filled with vapor.
Observing a tilted flask helps in deciding when all of the liquid has vaporized.

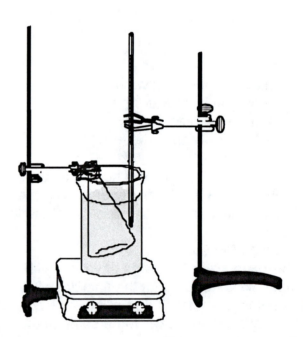

SAFETY

Wear safety glasses throughout this experiment. The chemicals used in this experiment are all volatile solvents, that is, liquids that evaporate readily. Because of their flammability, low-boiling liquids were chosen so that they could be vaporized using water baths, rather than an open flame.

Chemical	Toxicity	Flammability	Exposure
acetone (B.P. 56°C)	1	3	1
methanol (B.P. 65°C)	3	3	1
ethanol (B.P. 78°C)	2	3	1

0 is low hazard, 3 is high hazard

PROCEDURE

Part A: Molar Mass of a Volatile Liquid (acetone, methanol, or ethanol)

1. Cover a clean dry 125-mL Erlenmeyer flask with a piece of aluminum foil held in place with a rubber band.

2. Weigh the flask along with the foil cap to the nearest 0.01 g.

3. Remove foil. Using a small graduated cylinder, pour about 5 mL of liquid into the weighed flask. Replace foil and secure it with the rubber band. Make a single *very small* hole in the foil cover with a needle or pin.

4. Place about 450 mL of water in a 600-mL beaker. The water should be brought to a boil quickly then reduced to a gentle boil.   When the water begins to boil, adjust the hotplate setting so that the water remains boiling but does not splash out. Immerse the flask in the boiling water so that as much of the flask as possible is covered. Clamp flask to the ring stand so that it is tilted as shown in the figure in the apparatus section.

5. Excess vapor can be observed escaping from the hole in the foil.
Remove flask when you no longer see a Schlieren pattern emerging, that is, wavy lines seen rising from a heated surface. A paper towel or shiny surface can also be held over the hole. By keeping the flask slightly tilted, it will be easier to notice when the liquid has vaporized.

6. Allow flask to cool. (Liquid will condense.) Dry the surface with paper towel. Weigh flask plus contents to the nearest 0.01 g

7. Repeat two more times.

8. Choose another liquid and repeat the procedure, repeating trials as time allows.

Calculations

9. Find the volume (*V*) of the flask. Fill the flask to the very rim with tap water. Carefully pour the water from the flask into a graduated cylinder to determine the volume of the flask.

10. To find the mass of the liquid (*m*) requires the correction for lost air. Find the number of moles, *n*, of vapor using the ideal gas law. Here, the value of *T* is room temperature, *P* is the vapor pressure of the liquid compound at room temperature (See Table 13.1), and *V* is the volume of the flask. The number of moles of vapor is also the number of moles of lost air. Using the average molar mass of air, convert to grams and add to the mass of the cooled flask.

11. Find the mass of liquid (*m*) by subtracting the mass of the original "empty" flask from the corrected mass of the final cooled flask that was filled with vapor. Calculate *M* from the vapor density form of the ideal gas law. The *pressure* (*P*) is the barometric pressure in the room. The *temperature* (*T*) is the temperature of the boiling water bath (depends on atmospheric pressure but will be close to 100 or 373K).

Part B: Air Correction
For a given liquid, predict the importance of the correction for lost air. Check your prediction by calculating *M* with and without the correction for at least two of the volatile liquids provided (acetone, methanol or ethanol).

Data and Results (Molar Mass Liquid)

Name_____ Date_____ Section_____

Part A: Molar Mass of a Volatile Liquid

Liquid 1 _____   Room Temperature _____ °C

Vapor pressure _____ mm Hg   Barometric pressure _____ mm Hg

Finding Mass, m	Trial # 1	Trial # 2	Trial # 3
Mass of "empty" flask (g)			
Mass of flask plus vapor (g)			
Mass vapor (g)			
Air correction (PV = nRT)			
P vapor pressure @ room temp (atm)			
T Room temperature (K)			
V volume of flask (L)			
N liquid vapor (mol)			
n lost air (mol)			
mass lost air (g)			
Mass of flask plus vapor + lost air (g)			
Mass vapor, corrected (g)			
m average mass vapor (g)			

Finding Molar Mass, M	
m average mass vapor (g)	
P barometric pressure (atm)	
T Water bath temperature (K)	
V volume of flask (L)	
M molar mass, expt'l (g/mol)	
M molar mass, actual (g/mol)	
% error	

Instructor's Signature _____

Data and Results (Molar Mass Liquid)

Liquid 2 _____   Room Temperature _____ °C

Vapor pressure _____ mm Hg   Barometric pressure _____ mm Hg

Finding Mass, m	Trial # 1	Trial # 2	Trial # 3
Mass of "empty" flask (g)			
Mass of flask plus vapor (g)			
Mass vapor (g)			
Air correction (PV = nRT)			
P vapor pressure @ room temp (atm)			
T Room temperature (K)			
V volume of flask (L)			
N liquid vapor (mol)			
n lost air (mol)			
mass lost air (g)			
Mass of flask plus vapor + lost air (g)			
Mass vapor, corrected (g)			
m average mass vapor (g)			

Finding Molar Mass, M	
m average mass vapor (g)	
P barometric pressure (atm)	
T Water bath temperature (K)	
V volume of flask (L)	
M molar mass, expt'l (g/mol)	
M molar mass, actual (g/mol)	
% error	

Data and Results (Molar Mass Liquid)

Part B: Air Correction

Liquid	Vapor Pressure (mm Hg)	Molar Mass with Air Correction	Molar Mass without Air Correction

Questions

1. What advantage might there be in using a 250-mL or a 500-mL flask in this experiment?   What practical problem would be encountered by using a flask much larger than 125-mL?

2. Why must the flask that will be used to contain the vapor be weighed at least to the nearest 0.01 g?

3. Why were compounds such as ethylene glycol or methyl salicylate not among the liquids suggested for this experiment? (Refer to Appendix A.)

4. Show that the average molar mass for air is about 29 g/mol.

Experiment 14
Percent of H_2O_2 by Gas Evolved

Name_____ Date_____ Section_____

1. The decomposition of ammonium nitrite produces very pure nitrogen gas along with water. What mass of nitrogen gas can be prepared starting with 0.10 mole of ammonium nitrite? Why would nitrogen prepared by removing oxygen from air be less pure?

2. The total pressure of a mixture of nitrogen gas and water vapor is 758 mmHg. If the partial pressure of the water vapor is 26 mm Hg, what is the pressure of the nitrogen? What was the general principle you used to make the calculation?

3. A 500-mL flask contains pure helium gas. If the atmospheric pressure is 763 mmHg and the temperature is 20°C, what mass of helium is contained in the flask? What difficulties would be encountered in weighing the gas on a balance?

Instructor's Signature _____

Experiment 14
Percent of H₂O₂ by Gas Evolved

PURPOSE
The mass% of hydrogen peroxide present in a commercial product will be found by measuring the volume of O_2 produced upon decomposition using the enzyme, catalase, to increase the reaction rate. The effect of changing reaction conditions will be studied.

INTRODUCTION
Household hydrogen peroxide, found in concentrations of 2.5 to 3% in aqueous solution, is used medicinally for cleaning wounds, removing dead tissue, and even as a mild bleaching agent. Once a container of H_2O_2 solution is opened, the amount of peroxide gradually decreases as it decomposes to produce water and oxygen gas.

$$2H_2O_2(aq) \rightarrow 2H_2O\,(l) + O_2(g)$$

Concentrations of H_2O_2 can also be expressed in a unit called "volume strength". A 3% solution would be "10 volume" since one mL of solution releases about ten mL of oxygen at room temperature and atmospheric pressure.

The enzyme, catalase, increases the rate of decomposition of hydrogen peroxide by a factor of thousands. Catalase is found in most living cells and in the blood. A few tablespoons of household peroxide poured on a blood stain reacts with blood and removes it from carpets and clothing.

The major function of catalase within cells is to prevent the accumulation of toxic levels of hydrogen peroxide formed as a by-product of metabolic processes. It must be quickly converted into other, less dangerous, chemicals. Catalase has one of the highest turnover rates for all enzymes. One molecule of catalase can convert 100,000 molecules of hydrogen peroxide to water and oxygen each second.

Sources of catalase include yeast, horseradish root, liver and aqueous extract from potatoes. Dry yeast for baking, available in 7 g packages, is a particularly convenient source and will be used in this experiment.

The amount of hydrogen peroxide in the household product, roughly 3%, is determined by measuring the mass of oxygen produced. Since the oxygen is collected over water, the vapor pressure of water must be subtracted from the barometric pressure to find the pressure of oxygen. Vapor pressures of water at various temperatures are listed in Table 14.1.

Table 14.1 Vapor Pressure of Water

Temp (°C)	Vapor Pressure H_2O (mmHg)
18	15.5
19	16.5
20	17.5
21	18.7
22	19.8
23	21.1
24	22.4
25	23.8
26	25.2
27	26.7
28	28.3
29	30.0
30	31.8

APPARATUS

A setup that can be used for measuring the amount of oxygen gas by displacement of water is shown in the diagram. The reaction takes place in a large test tube and the oxygen is collected in a 50-mL graduated cylinder.

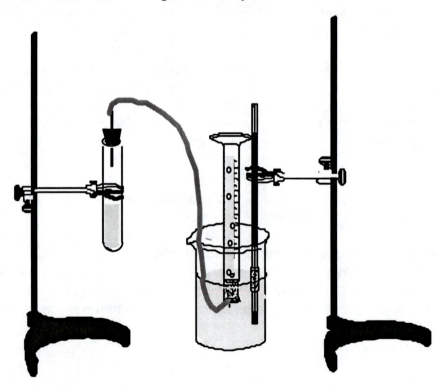

SAFETY

Wear safety glasses throughout this experiment.

Chemical	Toxicity	Flammability	Exposure
3% hydrogen peroxide (aq)	1	0	1
dry yeast	0	0	0
0.1 M hydrochloric acid	1	0	1
10% sodium hydroxide	2	0	2
1 M copper sulfate	1	0	1

0 is low hazard, 3 is high hazard

Concentrated (30%) hydrogen peroxide requires great care in handling and storing. When dropped on paper or wood, it can start a fire. Contact with the skin causes blotches that can be painful, but they disappear after a few hours without leaving traces.

PROCEDURE

Samples of hydrogen peroxide will be analyzed by measuring the total volume of oxygen gas produced. The rate of oxygen production with time is also followed. In Part B, the effect of pH and Cu^{2+} ions on the reaction will be examined.

Part A: Measuring Volume of O_2 from Decomposition of H_2O_2

1. Make a 10% slurry of dry yeast by mixing about 1 g dry yeast with 10 mL tap water.

2. Set up the gas collection device shown in the apparatus section using a 600-mL beaker, 50-mL graduate, and a large test tube with a one-hole stopper into which a short piece of tubing has been inserted. A length of flexible tubing connects the test tube to the graduate.

3. Fill the graduate with water, then invert into the beaker of water so that the graduate remains filled. (One way to do this is to cover the graduate with Parafilm, then remove it once the top of the graduate is submerged.) Make sure that the markings on the graduate are facing you.

4. Obtain a sample of the product labeled "3% H_2O_2" and record the date and time when the bottle was first opened.
Add the following to the test tube --- in the order below:

> 10 mL of tap water,
> 2 mL of the yeast slurry
> 4.0 mL H_2O_2 sample solution

Immediately reinsert the stopper in order to attach the reaction vessel to the gas-measuring device. Start a timer. Record the volume of O_2 produced vs. time. The first bubbles of oxygen should appear within 10 seconds.

5. When the reaction is over, that is, when the oxygen production stops, move the graduated cylinder vertically (keeping the open end submerged) until the water level inside the cylinder is the same as the water level in the beaker. This ensures that the total pressure on the gases inside the cylinder is the same as the barometric pressure in the room. Record gas volume trapped in the cylinder. Record the time required and the temperature of the water bath.

6. Use the ideal gas law to convert volume of oxygen to moles of oxygen, making sure to subtract the vapor pressure of water from the total pressure (See Table 14.1). Calculate the mass of hydrogen peroxide that decomposed and the %mass/volume (g H_2O_2/100 mL H_2O).

Part B: Changing Reaction Conditions

Observe how the catalase decomposition of H_2O_2 is affected when conditions are changed. Use reagents and equipment provided. All data/results must be recorded with units and when possible, should be tabulated and/or graphed.
To set up the reactions, add

> 10 mL of tap water
> 2 mL of the yeast slurry
> additive
> 4.0 mL H_2O_2 sample solution

The additives include the following reagents:
0.1 M HCl: try to reach pH 3 (about 1-2 mL or 25 drops)
10% NaOH: 15 drops
1 M $CuSO_4 \cdot 5H_2O$:   10 drops

1. Choose one of the three additives and observe what happens.
If time permits allow the reactions to run their course. Otherwise choose a stopping time and compare amount of oxygen produced with the amount in Part A.

Data and Results (Percent H₂O₂)

Name_____ Date_____ Section_____

Part A: Measuring Volume of O₂ from Decomposition of H₂O₂

H_2O_2 sample _____

Volume yeast slurry _____ mL

Volume water _____ mL

Volume "3%" H_2O_2 _____ mL

Volume O_2 produced _____mL

Total barometric pressure _____ mm Hg

Water bath temperature _____ °C

Vapor pressure H_2O _____ mm Hg

Pressure O_2 _____ mm Hg

moles O_2 _____ mol

moles H_2O_2 _____ mol

mass H_2O_2 _____ g

mass % H_2O_2 _____ %

Instructor's Signature _____

Data and Results (Percent H₂O₂)

Production of O_2 vs. Time

Time (min)	Vol O_2 (mL)

Part B: Changing Reaction Conditions

Tabulate your results.

Questions

1. Show that 3% hydrogen peroxide solution is close to "ten volume strength." Assume that the concentration is 3 g H_2O_2 per 100 mL water and that the density is about 1 g/mL. (See Introduction)

2. Why was the volume of the sample solution of hydrogen peroxide solution measured more accurately than the volume of water and the volume of yeast slurry?

3. Why is it necessary to equalize the levels of water in the graduated cylinder and the beaker before measuring the volume of gas collected?

4. Find the molarity of H_2O_2 in a 3.0% solution. Assume the density of the solution is 1.0 g/mL.

Experiment 15
Equilibrium and Le Châtelier's Principle

Name_____ Date_____ Section_____

1. Set up an equilibrium constant K in terms of molar concentrations for the synthesis of ammonia from its elements. Will the expression depend on whether you balance the reaction with whole numbers or fractions?

2. An organic compound called an ester is produced along with water by the reaction of an alcohol and an acid according to the equilibrium reaction:
$$\text{Alcohol} + \text{Acid} \rightleftharpoons \text{Ester} + H_2O$$
List all the ways in which the reaction could be driven to form more ester by changing the concentration of a reactant or product (other than adding ester).

3. A sealed tube is filled with an equilibrium mixture of dinitrogen tetroxide and nitrogen dioxide gases. When the tube is heated it becomes darker brown and when cooled it turns nearly colorless. For the decomposition of dinitrogen tetroxide:
a) Write the equilibrium reaction.
b) Is it exothermic or endothermic?
c) What piece of information is required to answer part b)?

Instructor's Signature _____

Experiment 15
Equilibrium and Le Châtelier's Principle

PURPOSE
The effect on equilibrium systems of changes in concentration and temperature will be examined using predictions from Le Châtelier's Principle.

INTRODUCTION
A brief description of chemical equilibrium and the famous principle of Henri-Louis Le Châtelier (1850-1936) follows. Le Châtelier's research included an effort to apply his own principle to the synthesis of ammonia. He also revolutionized the teaching of chemistry by emphasizing general principles rather than presenting students with lists of compounds and properties to be memorized.

Equilibrium
Chemical reactions that do not go to completion, where there is less than 100% conversion of reactants to products, are reversible. When the forward and reverse reactions are proceeding at the same rate, equilibrium is achieved. One classic example is the endothermic (heat is absorbed) synthesis of ammonia from its elements.

$$Heat + N_2(g) + 3H_2(g) \rightleftharpoons 2NH_3(g)$$

The successful method for reacting hydrogen with unreactive nitrogen to produce ammonia, that could be made into nitrogen compounds essential for growing plants, is one of the most important processes ever developed. It was Le Châtelier's Principle that made it possible.

LeChâtelier's Principle
A simplified version of LeChâtelier's Principle states that when conditions of an equilibrium system are changed, the equilibrium shifts to restore the original conditions. The three changes possible are described using the synthesis of ammonia as an example.
1. An increase in *pressure* will shift the equilibrium in the direction that gives fewer moles of gas. In the ammonia synthesis increasing the pressure would favor the production of product by driving the reaction toward the right producing 2 mol NH_3 compared to 4 mol reactant gases.
2. When the *concentration* of a reactant or product is changed, the equilibrium shifts to use up or make more of the reactant or product in question. For instance, removing ammonia from the reaction (actually done by liquefying the gaseous NH_3 as it forms) would favor the production of more ammonia.
3. What happens upon changing the *temperature* depends upon whether the reaction is exothermic (evolves heat) or endothermic (absorbs heat). Since the ammonia synthesis is endothermic, lowering the temperature would shift the equilibrium to the ammonia product. However, the temperature must be kept high enough to allow the reaction to proceed at a reasonable rate.

Synthesis of Ammonia
LeChâtelier tried to apply his ideas about stress and equilibrium to the ammonia synthesis using a pressure of 200 atm and the lowest temperature that would not inhibit the rate of the reaction. When the nitrogen and hydrogen were placed into a small steel container, air somehow entered the system. The container exploded and pieces of steel flew everywhere. Le Châtelier did not repeat the experiment. Less than five years later in 1912, the German chemist Fritz Haber did succeed in using Le Châtelier's ideas to synthesize ammonia for which Haber won the Nobel Prize in 1918.

In this experiment you will be working with solutions and solids so the stresses to be applied will include just two of the three----change in concentration of a species or change in temperature.

APPARATUS
Simple glassware such as test tubes, beakers and graduates along with a hot plate will be sufficient.

SAFETY
Wear safety glasses throughout this experiment. Use gloves when handling 12 M HCl (concentrated). Do not inhale vapors.

Chemical	Toxicity	Flammability	Exposure
saturated NaCl solution	0	0	0
saturated NH_4Cl solution	0	0	1
0.1 M $CaCl_2$	0	0	0
0.1 M Na_2CO_3	0	0	1
0.1 M $MgSO_4 \cdot 7H_2O$	0	0	1
0.1 M NaOH	1	0	1
12 M HCl	3	0	3
6 M HCl	2	0	2
1 M HCl	2	0	2

PROCEDURE

Part A: Saturated NaCl Solution
The solution labeled 'saturated NaCl' contains about 36 g NaCl in 100 mL water (the maximum that will dissolve at room temperature). The equation below describes the equilibrium to be examined:

$$NaCl(s) \rightleftharpoons Na^+(aq) + Cl^-(aq)$$

1. Place about 1 mL of the saturated NaCl solution in a test tube.

2. Add concentrated (12 M) HCl dropwise, until a change is observed. Record what happens.

⚠**CAUTION:** The 12 M HCl is concentrated. Do not inhale its vapors. Use gloves.

Part B: Saturated NH$_4$Cl Solution
In this part you will examine the equilibrium:

$$NH_4Cl(s) \rightleftharpoons NH_4^+(aq) + Cl^-(aq)$$

1. Using a 10-mL graduated cylinder, place about 1 mL (about 20 drops) of saturated NH$_4$Cl solution in a test tube. Increase the [Cl$^-$] by adding concentrated (12 M) HCl solution dropwise, until a change is observed. Record what happens.

2. Boil about 100 mL of tap water in a 250-mL beaker. Place the test tube with NH$_4$Cl and HCl into the boiling water. Shake and heat the tube for 3 minutes. Record what happens.

3. Obtain a qualitative value (positive or negative) for the heat of solution for NH$_4$Cl by noting what happens when about 2 mL of distilled water is added to 1-2 g of solid NH$_4$Cl. After shaking the test tube for a short time (in order to make dissolution faster), touch it with your bare hand. Is it cooler or warmer than it was before? Record your observation. Is this an endothermic or exothermic reaction?

Part C: Formation of the Copper Ammonium Complex

1. Place 2 mL of 0.1 M copper sulfate solution in a test tube.

2. Add 1 M NH$_3$ dropwise until a color change is observed.

3. Add 6 M HCl dropwise and note what happens.

Part D: Solubility of Calcium Carbonate

1. Mix 1 mL of 0.1 M calcium chloride solution with 1 mL of 0.1 M sodium carbonate solution in a test tube until a precipitate forms.

2. Add 6 M HCl dropwise until a change occurs. Describe what happens.

Part E: Solubility of Magnesium Hydroxide

1. Mix 1 mL of 0.1 M magnesium sulfate heptahydrate solution with 1 mL 0.1 M sodium hydroxide solution in a test tube until a precipitate forms.

2. Add 1 M HCl dropwise until a change occurs. Record result.

Data and Results (Equilibrium)

Name_____ Date_____ Section_____

Equilibrium Reaction	Stress Applied	Change(s) Observed*	Shift in Equilibrium and Reason

* Change in appearance (color or physical state), temperature, etc.

Instructor's Signature _____

Questions

1. Explain why the following substitutions in the procedure for this experiment could not be made:

a) potassium chloride for calcium chloride in Part D.

b) calcium sulfate hemihydrate for magnesium sulfate heptahydrate in Part E. (hemi means ½)

2. Calculate K_c for the equilibrium describing the dissolving of NaCl (See Part A).

3. What happens to K when the temperature of an endothermic reaction is

a) increased?   b) decreased?

4. What is the effect of adding a catalyst to an equilibrium reaction?

Experiment 16
Calorimetry: Measuring Heat of Neutralization

Name_____ Date_____ Section_____

1. There are 15 Calories (15 kcal) in one 4.0 g serving of table sugar, which is sucrose, $C_{12}H_{22}O_{11}$. Find the number of
a) cal/g   b) kJ/g   c) kJ/mol

2. A 27.8 g mass of copper metal at 100°C is added to 100 g of water at 20.0°C (specific heat = 4.18 J/g°C) and the final temperature is 22.0°C. What is the specific heat of copper? Does your answer seem reasonable? Explain.

3. A 0.250 g sample of solid potassium is dropped into a beaker containing 100 mL of water.
a) Give a balanced chemical equation for the reaction that takes place; one product is aqueous potassium hydroxide.
b) Calculate the heat of the reaction using $\Delta H°_f$ (in kJ/mol):$H_2O(l)$ -286   KOH(aq) -482
c) Is the reaction exothermic or endothermic?

Instructor's Signature _____

Experiment 16
Calorimetry: Measuring Heat of Neutralization

PURPOSE
The enthalpy change accompanying the neutralization of acids with NaOH(aq) will be measured using a polystyrene cup calorimeter.

INTRODUCTION
The heats of reaction will be measured for neutralizations including the reaction of sodium hydroxide with hydrochloric acid:

$$NaOH\ (aq)\ +\ HCl\ (aq)\ \rightarrow\ NaCl\ (aq)\ +\ H_2O\ (l)$$

Thermochemistry
Thermochemistry is the study of heat changes accompanying chemical reactions. The heat change for a reaction at constant pressure, change in enthalpy, ΔH_{rxn}, can be calculated from standard heats of formation, ΔH_f°, of products (P) and reactants (R). The sum of the heats of reaction of reactants are subtracted from those of the products where n_P and n_R refer to moles of product and reactant:

$$\Delta H_{rxn}^\circ = \sum_P n_P \Delta H_f^\circ (P) - \sum_R n_R \Delta H_f^\circ (R)$$

Standard heat of formation data needed to compare calculated heats of reaction to those measured in this experiment can be found in Table 16.1.

Table 16.1 Standard Heat of Formations

Substance	ΔH_f° (kJ/mol)
NaAc(aq)	-729.02
HAc(aq)	-485.8
NaOH(aq)	-470.1
NaCl(aq)	-407.3
H_2O(l)	-286.0
HCl(aq)	-167.2

Calorimetry

A simple calorimeter constructed from polystyrene cups, a thermometer and stirrer can be used to measure heats of reactions in solution. The calorimeter is open to the atmosphere so that the reaction takes place at approximately constant pressure. The heat absorbed or evolved is measured by noting the temperature change of a known mass of solution in the calorimeter. The heat capacity for the dilute solutions used is about the same as that of water or 4.18 J/g °C.

APPARATUS

A remarkably effective calorimeter is made from a thermometer and one or two polystyrene cups with a lid. The transparent cup in the diagram shows the space needed between the stir bar and the thermometer. A timer is provided to record time vs. temperature data.

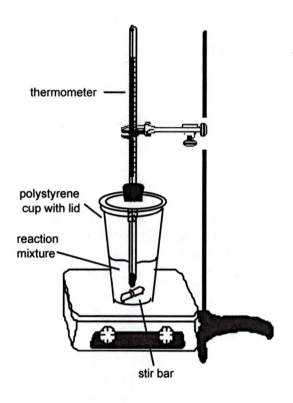

thermometer

polystyrene cup with lid

reaction mixture

stir bar

SAFETY

Wear safety glasses and gloves throughout this experiment.

Chemical	Toxicity	Flammability	Exposure
3.0 M HCl	2	0	2
3.0 M NaOH	2	0	2
0.90 M acetic acid	1	1	1
0.90 M NaOH	2	0	2

0 is low hazard, 3 is high hazard

PROCEDURE

The total heat change includes both the reaction and the calorimeter itself:

$$q_{total} = q_{reaction} + q_{calorimeter}$$

Very good results can be obtained using just the heat of the reaction as done in Parts A and B. In Part C the value of the calorimeter constant, C, is found and used to calculate $q_{calorimeter}$, the heat change for the calorimeter. The heat of reaction is then calculated again, this time taking into account $q_{calorimeter}$.

Part A: Heat of Reaction, NaOH + HCl

1. Place 50 mL of 3.0 M HCl solution into the calorimeter using a graduated cylinder, then replace the magnetic stir bar, thermometer and lid. While stirring slowly, allow the system to equilibrate for about five minutes.

2. Place 50 mL of 3.0 M NaOH solution into an Erlenmeyer flask.

3. Record and plot the temperatures of these solutions (T_A for acid and T_B for base) for a 3-4 minute period, taking measurements every minute. If the temperature of either solution changes, keep taking measurements until they are constant (or close).

4. Transfer the base solution from the flask to the calorimeter. Take measurements every 30 seconds. You will probably need at least 6 readings. Record temperatures and plot the data as shown in the diagram where the vertical line is the time when the cold and hot water are mixed. Extrapolate the linear portion of the plot back to the point just *after* mixing to find the final temperature, T_F.

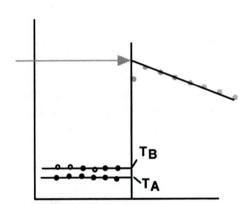

5. Repeat steps 1-4 to obtain a second measurement.

6. Calculate the heat of reaction by following the outline in the Data and Result sheet.

Part B: Heat of Reaction: NaOH + HAc
Repeat Part A using 0.90 M acetic acid and and 0.90 M NaOH in place of 3.0 M HCl and 3.0 M NaOH.

Part C: Calorimeter Constant (C_{cal})
The calorimeter itself gains a relatively small amount of heat. To find the calorimeter constant, C_{cal}, 50 mL of hot water is mixed with 50 mL of cold water. Starting temperatures are known. If the calorimeter worked perfectly, the final temperature calculated would be the same as the one measured. However, because the calorimeter absorbs heat, that will not be the case. The water and aqueous solutions have a

specific heat of 4.18 J/g °C and a density of 1.00 g/mL. The temperature right *after* mixing is the final temperature, T_F , T_H is the temperature of the hot water and T_C of the cold water. Both masses are 50.0 g. To find C_{cal} :

$$m_H \ s \ (T_H - T_F) \quad = \quad m_C \ s \ (T_F - T_C) \quad + \quad C_{cal}(T_F - T_C)$$

Heat lost	Heat gained	Heat gained
by hot water	by cold water	by calorimeter

From the change of temperature C_{cal} is calculated. The specific heat of water is s, m_H is the mass of hot water and m_C is the mass of cold water. The total volume of 100 mL will be the same in all the neutralization reactions to be run. The value of C_{cal} will also be the same.

1. Put 50 mL water in the calorimeter cup using a graduated cylinder, then replace the magnetic stir bar, thermometer and lid. While stirring slowly, allow the system to equilibrate for about five minutes.

2. Using a graduated cylinder, measure 50 mL of water into a 125-mL Erlenmeyer flask and heat it to about 60° C. (Use a second thermometer to check the temperature.) Remove the flask from the heat and allow it to equilibrate for about 5 minutes.

3. Record temperature vs. time for both the hot water and the cold water (in the calorimeter cup). You must do this simultaneously! Take three or four readings over a few minutes. Record on data sheet. Pour all of the warm water into the calorimeter. Replace the calorimeter lid and stir slowly using a stir bar while watching the temperature of the system.

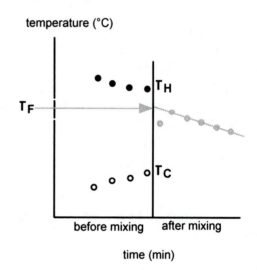

4. Record the temperature every 30 seconds. Five or six readings should be enough. To find the final temperature, T_F, extrapolate to the vertical mixing line as shown on the plot. Use the data obtained for T_H, T_C and T_F to find C_{cal}.

D: Calculation of Corrected Heat of Reaction for NaOH and HCl

1. To find the heat lost by the calorimeter --- the final temperature is T_F (just after mixing) and T_A is the initial temperature of the calorimeter, which contained the HCl solution.

$$q_{calorimeter} \ = C_{cal} \ (T_F \ - T_A)$$

2. The heat of reaction can now be recalculated using data from Part A.

$$q_{total} \ = \ q_{reaction} \ + \ q_{calorimeter}$$

Outline your calculation.

Data and Results (Heat of Neutralization)

Name_____ Date_____ Section_____

Part A: Heat of Reaction, NaOH + HCl

Trial 1

Time (min)	T_A (°C)	T_B (°C)	$T_{mixture}$ (°C)
Before mixing			
After mixing			

Trial 2

Time (min)	T_A (°C)	T_B (°C)	$T_{mixture}$ (°C)
Before mixing			
After mixing			

Instructor's Signature _____

Data and Results (Heat of Neutralization)

Calculating Heat of Reaction for NaOH + HCl

NaOH + HCl	Trial #1	Trial #2
$T_A =$	°C	°C
$T_B =$	°C	°C
$T_0 =$	°C	°C
$T_F - T_A =$	°C	°C
$T_F - T_B =$	°C	°C
$q_A = m_A \times s \; (T_F - T_A) =$	J	J
$q_B = m_B \times s \; (T_F - T_B) =$	J	J
$q_R = q_A + q_B$	J	J
n = mol of either reactant	mol	mol
$\Delta H = -q_R/n$	kJ/mol	kJ/mol

Mean of the two ΔH values   = _____ kJ/mol (NaOH + HCl)

Data and Results (Heat of Neutralization)

Part B: Heat of Reaction: NaOH + HAc

Trial 1

Time (min)	T_A (°C)	T_B (°C)	$T_{mixture}$ (°C)
Before mixing			
After mixing			

Trial 2

Time (min)	T_A (°C)	T_B (°C)	$T_{mixture}$ (°C)
Before mixing			
After mixing			

Data and Results (Heat of Neutralization)

Calculating Heat of Reaction for NaOH + HAc

NaOH + HAc	Trial #1	Trial #2
T_A =	°C	°C
T_B =	°C	°C
T_F =	°C	°C
$T_F - T_A$ =	°C	°C
$T_F - T_B$ =	°C	°C
$q_A = m_A \times s \ (T_F - T_A)$ =	J	J
$q_B = m_B \times s \ (T_F - T_B)$ =	J	J
$q_R = q_A + q_B$	J	J
n = mol of either reactant	mol	mol
$\Delta H = -q_R/n$	kJ/mol	kJ/mol

Mean of the two ΔH values   = _____ kJ/mol (NaOH + HAc)

Part C: Determination of the Heat Capacity of the Calorimeter (C)

Time (min)	T_C (°C)	T_H (°C)	$T_{mixture}$ (°C)
Before mixing			
After mixing			

Data and Results (Heat of Neutralization)

Calculation of the calorimeter constant:

(a) $T_F - T_c =$ _____ °C $T_H - T_F =$ _____ °C

(b) Heat lost by hot water $= m_H \times s\,(T_H - T_F) =$ _____ J

(c) Heat gained by cold water $= m_c \times s\,(T_F - T_c) =$ _____ J

(d) Heat gained by the calorimeter $=$ (b) - (c) $=$ _____ J

(e) $C_{cal} =$ (d) / $(T_F - T_c) =$ _____ J/°C   (C_{cal} should be positive)

D: Calculation of Heat of Reaction including $q_{calorimeter}$:

Questions

1. Standard heats of formation can be used to calculate heats of neutralization. Compare the calculated values with the mean of your measured values.

2. Compare the heat of reaction determined with and without the calorimeter constant.

3. Show that the same value for the heat of reaction can be found by averaging T_A and T_B to find q_R, rather than adding q_A and q_B.

4. In Part B why was 0.090 M acetic acid used instead of 3.0 M acetic acid?

Experiment 17
Heat of Solution and Hot Packs

___

Name_____ Date_____ Section_____

1. The heat of solution for sodium chloride is +3.9 kJ/mol and for sodium hydroxide it is -44.0 kJ/mol. Would heat be absorbed or evolved upon making an aqueous solution of
a) NaCl?   b) NaOH?

2. Suppose the heat evolved upon dissolving 0.400 g sodium hydroxide in water is 4,400 J. What is the heat of solution for sodium hydroxide in kJ/mol?

3. Find the mass % of water in calcium chloride dihydrate.

Instructor's Signature _____

Experiment 17
Heat of Solution and Hot Packs

PURPOSE
Heat increases will be measured for the dissolving of calcium chloride, the basis of commercial hot packs.

INTRODUCTION
Many hot packs rely upon the exothermic process of solution.

Heat of Solution and Hot Packs
The heat of solution for a substance is the heat given off or absorbed when one mole is completely dissolved in a large volume of water (meaning to infinite dilution). Heats of solution for compounds available from household products are listed below in Table 17.1. When the temperature increases upon mixing a solute and solvent heat is given off, the process of solution is *exothermic* and the heat of solution, $\Delta H_{sol'n}$, will be negative.

Table 17.1 Heats of solution of available anhydrous compounds

Compound	Formula	Heat of Solution (kJ/mol)
ammonium chloride	NH_4Cl	+14.8
sodium chloride	$NaCl$	+3.9
sodium hydroxide	$NaOH$	-44.0
calcium chloride	$CaCl_2$	- 81.3
magnesium sulfate	$MgSO_4$	-91.2

Many commercial hot packs are made from anhydrous $CaCl_2$. A small inner bag containing water is put inside of another bag containing a fixed amount of the salt. When the inner bag is crushed, the water and solid are mixed and the temperature rises. For a hot pack made from 25 g calcium chloride and 117 mL water, the temperature will rise from room temperature to 50°C. Doubling the mass of $CaCl_2$ while using the same amount of water gives an increase to 70°C.

APPARATUS

A simple calorimeter for measuring the heat of solution is made from a thermometer and a polystyrene cup. The thermometer must be immersed in liquid, but cannot touch the bottom of the cup. To keep the lightweight plastic cup from toppling over, it is supported in a beaker as shown. Careful swirling of the polystyrene cup can be used to mix water and salt.

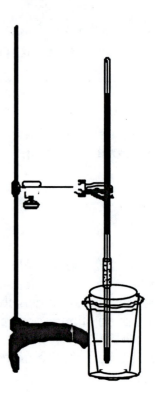

SAFETY

Wear safety glasses throughout this experiment.

Chemical	Toxicity	Flammability	Exposure
calcium chloride	1	0	1
ammonium chloride	1	0	1
sodium chloride	0	0	0
sodium hydroxide	3	0	3
magnesium sulfate	0	0	1

0 is low hazard, 3 is high hazard

PROCEDURE

Part A: Dehydration of Calcium Chloride Dihydrate
This is necessary only if the calcium chloride available is the dihydrate, rather than anhydrous.

1. Weigh a sample of calcium chloride dihydrate into a Petri dish.

2. Heat using a medium setting on a hot plate until the sample has lost 20 to 24% of its mass. The product is anhydrous $CaCl_2$.
⚠**CAUTION:** If the temperature is set too high, the Petri dish can break.

Part B: Heat Changes on Dissolving Calcium Chloride in Water
Different proportions of calcium chloride and water are used to duplicate some commercial hot packs.

1. Measure 50 mL of tap water into a polystyrene cup supported in a beaker. Position the thermometer as shown in the figure, so that the bulb is immersed in the water, but does not touch the bottom of the cup.

2. Weigh 5.0 g of anhydrous $CaCl_2$. An amount between 4.96 g and 5.04 g is fine. Record the initial temperature of the water to the nearest half-degree, $\pm 0.5°C$. Add the salt to the water--- all at once--- and start the timer. Swirl the flask to hasten the dissolving.

3. Record the temperature ($\pm 0.5°C$) when the increase or decrease stops, that is, when the temperature reaches a maximum or a minimum. This should take no more than two minutes, possibly less than one.

4. Pour used solution into a labeled bottle. Rinse the beaker with water, then repeat Steps 2 through 4. Record the starting temperature each time, since it may change if there are drafts in your room.

Part C: Other Candidates for Hot Packs
One other compound listed in Table 17.1 is used in commercial hot packs. Decide which one it is most likely to be. Measure its temperature change/mass. If the compound chosen is hydrated, calculate the mass % of water it contains, then remove the water as done in Part A for calcium chloride.

Data and Results (Hot Packs)

Name_____ Date_____ Section_____

Part A: Dehydration of Calcium Chloride Dihydrate

Mass of hydrated calcium chloride sample _____ g

Mass of water in the sample above _____ g

Mass anhydrous calcium chloride obtained _____ g

Percent water removed _____ % (should be 20-24%)

Part B: Heat Changes on Dissolving Calcium Chloride in Water

Trial	Mass CaCl$_2$ (g)	Vol. H$_2$O (mL)	Initial Temp. (˚C)	Maximum Temp. (˚C)	Increase in Temp. (˚C)
# 1					
# 2					

Part C: Other Candidates for Hot Packs

Instructor's Signature _____

Questions

1. Show that heating to remove 20 to 24% of the mass of a sample of calcium chloride dihydrate should leave a sample that is close to anhydrous. Approximately what % of magnesium sulfate heptahydrate must be removed to produce anhydrous $MgSO_4$?

2. For the commercial hot pack made from 25 g calcium chloride and 117 mL water, comment on the magnitude of temperature increase that occurs upon doubling the mass of $CaCl_2$ while using the same amount of water. Would you expect the increase to be exactly doubled? Explain.

Anhydrous $CaCl_2$ (g)	Initial Temp. (˚C)	Final Temp. (˚C)
25	24	51
50	23	70

3. Why would NaOH be a poor choice for use in commercial hot packs?

Experiment 18
Molar Mass by Freezing Point Depression

Name_____ Date_____ Section_____

1. Find the freezing point of an antifreeze solution made by mixing 500 g ethylene glycol, $C_2H_6O_2$ with 1000 g water. (k_f water = 1.86°C/m). Would this solution be effective throughout a cold winter?

2. Suppose 15 g of a carbohydrate, known to be either sucrose or ribose, is dissolved in 100 g water to make a solution that freezes at -1.9°C. Is it sucrose ($C_{12}H_{22}O_{11}$) or ribose ($C_5H_{10}O_5$)? What experimental problems could be encountered in using freezing point depression to find the molar mass of a compound such as sucrose?

3. For the ionic compounds below give the maximum number of ions that could be produced when they dissolve in water.
a) sodium chloride   b) potassium sulfate   c) calcium chloride

Instructor's Signature _____

Experiment 18
Molar Mass by Freezing Point Depression

PURPOSE
The freezing point depressions of aqueous solutions, such as urea, will be used to determine their molar masses.

INTRODUCTION
The freezing point of a solution is always lower than the freezing point of the pure solvent. The difference between the freezing point T_f of a pure solvent and that of a solution is the freezing point depression, ΔT_f :

$$\Delta T_f = i \times k_f \times m$$

For water k_f is 1.86 °C/m where m is the molality, mol solute/kg solvent. The value of i is the number of "units" formed when the compound dissolves. For nonionic compounds such as urea and most other organic compounds, i is 1. For ionic compounds the maximum value of i is the number of ions formed per formula unit. For instance, for every NaCl that dissolves, two ions, sodium and chloride, are produced. Thus, for NaCl the maximum "limiting value" for i is 2. Since 100% ionization takes place only in very dilute solutions, i decreases as the salt concentration increases. For a 0.0100 m NaCl solution i is 1.94 compared to 1.87 for a 0.100 m NaCl solution.

Since freezing is difficult to observe by eye, a temperature vs. time graph called a *cooling curve* is used instead. Freezing point depression can be used to find the molar mass of a solute by calculating its molality while knowing the mass of the solute and solvent used to prepare the solution.

Conversely, from knowing the molar mass of a compound and its molality, the freezing point depression can be found. For ionic compounds the value of i must be taken into account.

Antifreeze
Antifreeze is a substance added to water in the cooling system of an internal combustion engine to lower its freezing point so that it can be cooled below the freezing point of pure water without freezing. Ethylene glycol is the most widely used automotive antifreeze, although other alcohols can be substituted (See Table 18.1). An advantage of propylene glycol is its lower toxicity to humans and to pets who happen to like the sweet taste of ethylene glycol.

Table 18.1 Compounds used as antifreeze

Compound	Formula	Molar Mass (g/mol)
Methanol	CH_4O	32
Ethanol	C_2H_6O	46
Isopropyl alcohol	C_3H_8O	60
Ethylene glycol	$C_2H_6O_2$	62
Propylene glycol	$C_3H_8O_2$	76

APPARATUS

There must be enough room to insert the thermometer into the beaker so that the bulb is immersed in the liquid, but does not interfere with the stir bar. Small (30-mL) beakers will suffice, and are recommended to reduce the amount of material needed. Larger beakers can be used provided the liquid level is high enough so that the thermometer is submerged.

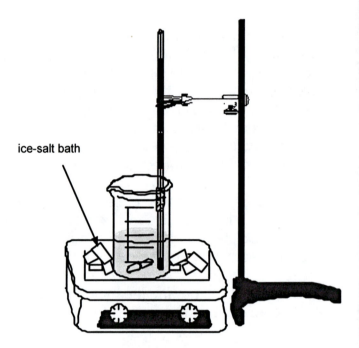

ice-salt bath

SAFETY

Wear safety glasses and gloves throughout this experiment. Waste containers will be available for all used solutions.

Chemical	Toxicity	Flammability	Exposure
urea	0	0	1
glycerol	0	1	1
ethylene glycol	2	1	2
KCl	1	0	1

0 is low hazard, 3 is high hazard

PROCEDURE

A 30-mL beaker is recommended so that the height of liquid is sufficient to have the thermometer bulb covered with liquid while limiting the amount of solute needed. If using a larger beaker, increase all amounts accordingly.

Part A: Freezing Point of Pure Water

1. Using the volume markings on the beaker and a water wash bottle, measure 15 mL water into a clean dry 30-mL beaker. Record the volume and mass of water on the data sheet.

2. Place the beaker of water in a wide dish which will be used for the cooling bath. Put this assembly on a magnetic stir plate, place the small stir bar in the beaker of water and begin stirring. Immerse the thermometer into the water between the stir bar and the side of the beaker. The bulb should be close to the bottom of the beaker, without touching it. See the diagram.

3. Prepare an ice-salt mixture by packing ice (about the amount that fits in a polystyrene cup) in the dish around the beaker. Add about 25 g salt being very careful not to get any in the beaker.

4. Let the water cool to about 8°C , then start measuring the temperature at least to the nearest 0.5°C (you may be able to estimate to 0.2 °C) every 30 seconds until the temperature becomes constant for several minutes. Record time and temperature data. It is not necessary to freeze the entire sample. "Supercooling" may occur, meaning that the temperature falls below the freezing point, then climbs back up before it remains constant. Stirring usually prevents this, but it may not.

5. Prepare a plot of temperature vs. time called the "cooling curve". The freezing point is the temperature at which the cooling curve levels off. The freezing point may not be 0°C. Since you will be measuring a difference in freezing points and using the same setup for both parts of the experiment, this will not make any difference.

6. Save the water sample setup with stir bar for the next part. Carefully lift the beaker out of the ice bath and let it warm up until any ice crystals melt. Remove the ice bath from the stir plate and drain away some of the excess water.

Part B: Freezing Point of Urea Solution

1. Weigh out 1.5 to 2.0 grams of urea using weighing paper and the top loading balance. You can weigh the urea to the nearest 0.01 g. Any mass in the range given will be fine. Record all three figures.

2. Carefully transfer the urea from the weighing paper into the same water you used for Part A. Put beaker with solution back on the stir plate and begin stirring. When the urea is completely dissolved, you can place the beaker into the ice bath and add more ice and salt as needed.

3. Carefully lower the thermometer into the solution and record the cooling curve data as above. Prepare a cooling curve and find the freezing point of the mixture.
If the cooling is slow, just pour more salt over the ice.

Part C: Freezing Point of Solutions of Glycerol and/or Ethylene Glycol
Find the molar mass of one or both of the liquids, ethylene glycol and glycerol, as time allows. Both are miscible with water. You can use the water cooling curve from Part A. Before starting the experiment have your instructor check your procedure which should include the amounts of liquid to be mixed with water as well as the plan for making the solution.

Part D: Freezing Point of Solutions of KCl
The source of KCl is a salt substitute product. Using the same procedure from Part B with similar quantities, prepare a cooling curve and find the freezing point depression of a KCl solution. Estimate the value of i for the solution used and from i, the degree of ionization for that concentration of KCl.

Record amounts used and attach the cooling curve. Give the reasoning you used to estimate the degree of ionization of KCl.

Data and Results (Freezing Point Depression)

Name_____ Date_____ Section_____

Part A: Freezing Point of Pure Water

Volume of water _____ mL

Mass of water used _____ g   = _____kg

Time (min)	Temp (°C)	Time (min)	Temp (°C)	Time (min)	Temp (°C)

Part B: Freezing Point of Urea Solution

Mass of urea _____ g

Time (min)	Temp (°C)	Time (min)	Temp (°C)	Time (min)	Temp (°C)

Instructor's Signature _____

Data and Results (Freezing Point Depression)

Freezing point of pure water (from the cooling curve), $T_f(H_2O)$ _____ °C

Freezing point of urea solution (from cooling curve), T_f(urea sol'n) _____ °C

Freezing point depression, $\Delta T_f = T_f(H_2O) - T_f$(urea solution), ΔT_f _____ °C

molality of solution: $\Delta T_f = k_f \times m$, therefore: $m = \Delta T_f / k_f$ m _____ mol/kg H_2O

(K_f for water = 1.86 °C /m)

Mass of water used (see previous page) _____ kg

Moles of urea: mol = kg H_2O x m   _____ mol

Mass of urea used (see previous page   _____ g

Molar mass of urea _____ g/mol   *Retain only 2 significant figures here.*

Molar mass of urea, CH_4N_2O, known _____ g/mol

Molar mass of urea measured   _____ g/mol

% Error = $(M_{measured} - M_{known}) / M_{known}) \times 100\%$ _____ %

Attach the cooling curves used for freezing point determinations

Data and Results (Freezing Point Depression)

Part C: Freezing Point of Glycerol Solution

Mass of glycerol: _____ g

Time (min)	Temp (°C)	Time (min)	Temp (°C)	Time (min)	Temp (°C)

Freezing point of pure water (from the cooling curve), $T_f(H_2O)$ _____ °C

Freezing point of glycerol solution (from cooling curve),
T_f (glycerol solution) _____ °C

Freezing point depression, $\Delta T_f = T_f(H_2O) - T_f$ (glycerol solution) ΔT_f _____ °C

molality of solution: $\Delta T_f = k_f \times m$, therefore: $m = \Delta T_f / k_f$ m _____ mol/kg H_2O

  (K_f for water = 1.86 °C /m)

Mass of water used (see previous page) _____ kg

Moles of glycerol = kg H_2O x m   _____ mol

Mass of glycerol used (see previous page) _____ g

Molar mass of glycerol _____ g/mol *Retain only 2 significant figures here.*

Molar mass of glycerol, $C_3H_8O_3$, known _____ g/mol

Molar mass of glycerol measured   _____ g/mol

Data and Results (Freezing Point Depression)

% Error = $(M_{measured} - M_{known}) / M_{known}$ x 100% _____ %

Attach the cooling curves used for freezing point determinations

Part D: Freezing Point of Solutions of KCl

Record amounts of KCl and water used, attach cooling curve and describe the method used to estimate the ionization of the KCl solution used.

Questions

1. Why must you retain no more than two significant figures in the molar masses determined from this experiment?

2. Sketch a cooling curve for pure water that would result if supercooling had taken place.

3. Of the compounds used in antifreeze, which one would produce the largest freezing point depression per gram of compound dissolved in a given amount of water?

4. Suppose a 1.0 m aqueous solution of an organic compound freezes at -0.93°C. What explanation would be consistent with this result? Recall that k_f for H_2O = 1.86°C/m.

Experiment 19
Kinetics of the Iodination of Acetone

Name_____ Date_____ Section_____

1. Using the data below for the reaction:

$$2NOCl(g) \rightarrow 2NO(g) + Cl_2(g).$$

Find the order with respect to NOCl

Experiment	[NOCl]$_o$ (M)	Initial Rate(M/s)
1	0.30	0.36×10^{-8}
2	0.60	1.44×10^{-8}
3	0.90	3.24×10^{-8}

2. For the reaction and data above, give the rate equation and find the value of the rate constant, k.

3. Could you have simply used the stoichiometry of the reaction to determine the order with respect to NOCl? Explain.

Instructor's Signature _____

Experiment 19
Kinetics of the Iodination of Acetone

PURPOSE
The rate law for the reaction of iodine with acetone will be found using the method of initial rates. The initial rate of each reaction will be determined by measuring the time required for the color of iodine to disappear.

INTRODUCTION
The rate of the iodination of acetone could depend on the concentrations of the reactants, acetone and iodine and the acid catalyst, HCl.

$$H_3C-\overset{\overset{O}{\|}}{C}-CH_3 \ + \ I_2 \quad \xrightarrow{\ H^+_{(aq)}\ } \quad H_3C-\overset{\overset{O}{\|}}{C}-CH_2I \ + \ HI$$

The rate of this reaction takes the form below, where 'a' is the order with respect to (wrt) acetone, 'i' , the order wrt iodine and 'h' the order for HCl.:

$$\text{Rate of Reaction} = k \, [\text{acetone}]^a[\text{iodine}]^i[\text{HCl}]^h$$

To complete the rate expression, a, i, and h can be determined using the method of initial rates. From the initial rate for a given set of concentrations, the rate constant, k, can be calculated.

Initial rates are measured for several reactions, while varying the concentration of one reactant at a time. For example, suppose the concentration of one reactant is doubled. If the initial rate also doubles, the reaction must be 1st order with respect to that reactant. If instead, the initial rate quadruples when the concentration doubles, the reaction must be 2nd order in that reactant.

For the iodination of acetone, six reactions will be run. The initial rate is measured with the assistance of a spectrometer by noting the time required for the brownish colored iodine to disappear. For each of the three reactants there will be at least one pair of reactions in which only that reactant varies while the other two remain constant. The order with respect to that reactant can thus be found.

APPARATUS
A spectrometer such as the Spectronic-20 is preset to 410 nm where the iodine absorbs.

SAFETY
Acetone is a flammable solvent. Wear safety glasses and gloves throughout this experiment.

Chemical	Toxicity	Flammability	Exposure
0.0050 M iodine	0	0	1
4.0 M acetone	1	3	1
1.0 M HCl	2	0	2

0 is low hazard, 3 is high hazard

PROCEDURE
Noticing the disappearance of one reactant, the purplish I_2, will be used to follow the reactions.

Part A: Running the Reactions
Six reactions are run using varying amounts of reactants. The table below gives sets of concentrations that can be used. The order in which the reactions are run does not matter.

Reaction No.	4.0 M Acetone (mL)	1.0 M HCl (mL)	Water (mL)	0.0050 M Iodine (mL)
1	3.00	3.00	8.00	4.00
2	6.00	3.00	5.00	4.00
3	9.00	3.00	2.00	4.00
4	3.00	6.00	5.00	4.00
5	3.00	9.00	2.00	4.00
6	3.00	3.00	4.00	8.00

1. The spectrometer is set at a wavelength of 410 nm. Have the instrument calibrated for this wavelength, using a cuvet (or test tube) filled with water. Remove the cuvet and empty the water. Collect about 30 mL of each solution in a clean beaker and cover with Parafilm.   Record the room temperature.

2. It is important to be ready to put the sample cuvet into the spectrometer as soon as possible after the reactants are mixed. One graduated pipet should always be used for the iodine solution and the other for the acetone, acid and water. After pipeting the acetone, rinse first with HCl and then pipet the HCl. Do the same for water.

For Reaction # 1:
Pipet into one beaker: 3.00 mL acetone , 3.00 mL HCl , 8.00 mL water
Into another beaker, pipet 4.00 mL iodine
To start the reaction, pour the iodine solution into the beaker containing the acetone, acid and water. Mix as quickly as you can and set the timer (Two people should work together on this—one mixing and one timing.) Fill the cuvet about ¾ full, place in the spectrometer and observe the %T, which will increase as the iodine reacts. When the increase stops, the reaction is over. Record the time.

Repeat the reaction until the times agree within 10% of one another.

3. Perform reactions 2 through 6.

4. Adjust the concentrations of acetone, acid and iodine for dilution, then use the data to find the rate of reaction [iodine]/time(s)

5. Find the order with respect to each reactant, a, h, and i, explaining how each was obtained. Write the rate law. Calculate the rate constant, k, for each set of data, then find the average.

Part B: A 7th Reaction
Find the concentrations needed for one more reaction that confirms the order with respect to iodine. Record the room temperature. Perform the reaction and analyze the data.

Data and Results (Iodination of Acetone)

Name_____ Date_____ Section_____

Part A: Running the Reactions

Temperature _____°C

Reaction No.	4.0 M Acetone (mL)	1.0 M HCl (mL)	Water (mL)	0.0050 M Iodine (mL)	Time (s)
1	3.00	3.00	8.00	4.00	
2	6.00	3.00	5.00	4.00	
3	9.00	3.00	2.00	4.00	
4	3.00	6.00	5.00	4.00	
5	3.00	9.00	2.00	4.00	
6	3.00	3.00	4.00	8.00	

Corrected concentrations, rate in $[I_2]$/s, and rate constant

Reaction No.	Acetone (M)	HCl (M)	Water (mL)	Iodine (M)	Rate ($[I_2]$/s x 10^5)	Rate Constant (k x 10^5)
1			8.00			
2			5.00			
3			2.00			
4			5.00			
5			2.00			
6			4.00			

Instructor's Signature _____

Data and Results (Iodination of Acetone)

Order wrt acetone _____

Order wrt HCl _____

Order wrt iodine _____

Rate Law

Part B: A 7th Reaction

Temperature _____ °C

Questions

1. The coefficient of iodine in the reaction studied is 1, however, the order with respect to I_2 is zero. Explain.

2. What problem could arise if the reactions are not all done on the same day?

3. Why must reactant solutions be covered before they are mixed?

4. Why can the reactions be done in any order?

Experiment 20
Kinetics of the Oxidation of Food Dye with Sodium Hypochlorite

Name_____ Date_____ Section_____

1. The concentration vs. time data below was collected for the decomposition of N_2O_5 at 55°C. $2 N_2O_5(g) \rightarrow 4 NO_2(g) + O_2(g)$

Time (s)	0	100	200	300	400	500	600	700
$[N_2O_5]$(M)	0.020	0.011	0.014	0.012	0.010	0.0086	0.0070	0.0061

What quantity should be plotted vs. time to find out if the order with respect to N_2O_5 is:
a) 0th ?   b) 1st ?   c) 2nd ?

2. Make a plot to show that the decomposition of N_2O_5 is first order. Write the rate equation and find the value of the rate constant, k?

3. For first order reactions the half-life, $t_{1/2} = 0.693/k$, is independent of concentration. How could you show that the reaction is first order with respect to N_2O_4 by simply inspecting the data given in Number 1? What is the half-life and the value of k?

Instructor's Signature _____

Kinetics of the Oxidation of Food Dye with Sodium Hypochlorite

PURPOSE

The rate law for the reaction of bleach with food dye will be found by measuring the disappearance of color to find the order with respect to each reactant.

INTRODUCTION

Sodium hypochlorite (bleach) is an oxidizing agent that can decolorize stains on fabrics by reacting with colored substances. Often (not always) the oxidized form of the stain is less colored than the untreated form. If so, the stain disappears.

$$\text{Colored substance} + \text{NaOCl} \rightarrow \text{Colorless products}$$

In this experiment bleach reacts with the blue food coloring shown. One feature of many colored compounds such as this, is the presence in their formulas of alternating single and double bonds. Oxidizing agents react with the double bonds.

Blue No. 1
Brilliant Blue

Reaction Rate

The rate of this reaction takes the form below, where 'b' is the order with respect to sodium hypochlorite (bleach) and 'n' is the order with respect to the food dye.

$$\text{Rate of Reaction} = k\,[\text{NaOCl}]^b[\text{food dye}]^n$$

Using a large excess of bleach so that its concentration is essentially unchanging, the rate law becomes:

$$\text{Rate of Reaction} = k'\,[\text{food dye}]^n$$

The concentration of bleach is incorporated into the observed rate constant, k':

$$k' = k[\text{NaOCl}]^b$$

The reaction is said to follow pseudo nth order kinetics.

Blue No.1 has a maximum at 630 nm. To determine the value of n, the disappearance of color upon reaction with bleach is followed by using a Spectronic-20 set to 630 nm along with a timer, or a computer-interfaced colorimeter with data collecting software.

To find n, three simple rate laws are tested by looking for a straight line upon plotting A (0th order), ln A (1st order) or 1/A (2nd order) vs. time. From the value of n, k' can be found. To determine k, the order with respect to bleach (b) must be known as well. The value of b is found by running the same reaction, once again with an excess of bleach, but this time using a *different* excess. What happens to the observed rate constant, k', reveals the order with respect to bleach. For instance, if k' is doubled when [NaOCl] is doubled, then b must be 1.

APPARATUS

A spectrometer set to 630 nm can be used to follow the disappearance of blue color.

A colorimeter interfaced with a computer may also be used at 635 nm (one of the available wavelengths).

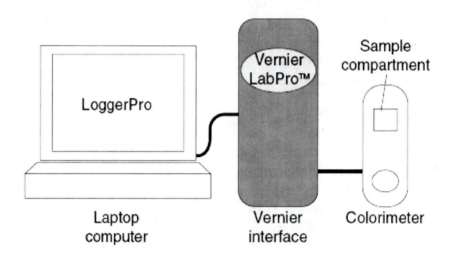

SAFETY

Wear safety glasses and gloves throughout this experiment. Commercial sodium hypochlorite is a strong oxidizing agent that should be handled with care. Do not let it touch your skin or clothing. If some is spilled, clean the area with plenty of water.

Chemical	Toxicity	Flammability	Exposure
food dyes	0	0	0
5-6% NaOCl (aq)	2	0	2

0 is low hazard, 3 is high hazard

PROCEDURE

Two different sets of equipment can be used to do this experiment. Parts A and B describe a spectrometer with timer. Parts C and D use a colorimeter interfaced to a computer with a Vernier interface.

Part A: Order with Respect to Dye (Spectrometer)

1. Dilute the dye by adding one drop of food coloring using a Pasteur pipet into a 100-mL volumetric flask then diluting with water to the mark. This should put the maximum absorbance between 0.5 and 1.2.

2. Set the wavelength to 630 nm. Zero the spectrometer using a water blank.

3. A graduate can be used to measure volumes. Add 25 mL food dye solution to a 50-mL beaker. Add 1.0 mL bleach to the dye solution and swirl to mix. If most of the color disappears in less than five minutes, the bleach solution should be diluted to slow down the reaction.

4. Use the bleach solution that provides a reaction time of at least 10 minutes (Step 3), Working *as quickly as you can*, mix the food dye and bleach, transfer reaction mixture to fill the cuvet, place in the spectrometer and take a %T reading immediately.
(The progress of the reaction mixture left in the beaker can also be followed visually.)

5. Continue to record the %T at 630 nm every minute (or every 30 s) until the %T is greater than 90%.

6. Plot the quantities A, ln A and 1/A vs. time to find the order of the reaction with respect to the dye.

Part B: Order with Respect to Bleach (Spectrometer)

1. Dilute the bleach solution used in Part A by half using 1.0 mL bleach diluted with distilled water to 2.0 mL

2. Once again mix the bleach and food dye and follow the disappearance of A at 630 nm. Comparing the slope from the result in Part A will reveal the order with respect to bleach.

Part C: Order with Respect to Dye (Colorimeter/Computer)

1. Prepare solutions as done in Part A (Steps 1-3).

2. *Set-Up.* Connect the colorimeter sensor to the channel 1 port of the Vernier interface. Attach the interface to the back of the computer using the usb cord. Open LoggerPro. Click on LabPro. Choose Colorimeter from the analog sensors and drag the sensor to channel 1.

3. *Calibration*
Click on the Colorimeter sensor, select Calibrate and click on Calibrate Now.
Set Colorimeter selector knob to 0% T. Enter '0' and press Keep.
Set the wavelength on the colorimeter device to 635 nm. The molar absorptivity at 635 nm is also $1.38 \times 10^5 \, M^{-1}cm^{-1}$. (The fact that the maximum is 630 nm makes no difference in this experiment.) Open the colorimeter lid and insert a cuvette with water for the blank (100%T). Enter '100' and press Keep. Click on Done. Remove the blank cuvette. The sensor is now calibrated.

4. *Data Collection*
Fill a cuvette with the food dye solution and the Absorbance (for time 0) will be displayed. Remove cuvette. In the sampling rate section (clock icon), enter 2 samples per minute. Pipette 25.0 mL food dye solution into a 50-mL beaker. Mix 1.0 mL bleach solution with the dye solution. Swirl to mix. *Working quickly*, transfer reaction mixture to fill a clean cuvette and place in the holder. Press Collect. Press Stop when the absorbance reading falls below 0.1 A or %T > 90.

5. Record the Absorbance and time data points displayed on the computer screen. Plot quantities A, ln A, 1/A vs. time to find the order of the reaction with respect to the dye.

Part D: Order with Respect to Bleach (Colorimeter Computer)

1. Dilute the bleach solution used in Part A by half using 1.0 mL bleach diluted with distilled water to 2.0 mL

2. Once again mix the bleach and food dye and follow the disappearance of A at 635 nm. The same quantity that produced a straight line in Part C is plotted vs. time. Comparing the slopes of the two plots will reveal the order with respect to bleach.

Data and Results (Bleach)

Name_____ Date_____ Section_____

Part A: Order with Respect to Dye (Spectrometer)

Dilution factor for bleach _____

Time (min)	%T	A	ln A	1/A

Order with respect to dye _____

Instructor's Signature _____

Data and Results (Bleach)

Part B: Order with Respect to Bleach (Spectrometer)

For reaction with bleach diluted by half:

Time (min)	%T	A	ln A	1/A

Order with respect to bleach _____

Data and Results (Bleach)

Part C: Order with Respect to Dye (Colorimeter/Computer)

Dilution factor for bleach _____

Time (min)	A	ln A	1/A

Order with respect to dye _____

Data and Results (Bleach)

Part D: Order with Respect to Bleach (Colorimeter/Computer)

For reaction with bleach diluted by half:

Time (min)	A	ln A	1/A

Order with respect to bleach _____

Questions

1. Find the concentration of 6.0 % Clorox®, NaClO (aq). Assume that the density of the solution is 1.0 g/mL.

2. Calculate the concentration of a solution of blue dye #1 which has an absorbance of 1.0. The path length is 1 cm and the molar absorptivity, ε, is 1.38×10^5.

3. Suppose 1.0 mL of bleach is added to 25.0 mL of the blue dye solution above (#2). Show that the bleach is present in large excess. You can assume that the volumes are additive.

4. What was the order of the dye-bleach reaction with respect to the blue dye? How was this determined?

5. What was the order of the dye-bleach reaction with respect to bleach? How was this determined? Write the rate equation for the dye-bleach reaction.

6. Find the value of the pseudo order rate constant, k', from the slope of the plot used to determine reaction order with respect to the dye. What is the value of k?

7. Could the method used in this experiment be applied to a reaction that is over within a minute? Why?

Experiment 21
Redox Titration:
Iodine Used to Determine Ascorbic Acid

___

Name_____ Date_____ Section_____

1. Assign oxidation numbers to iodine in the following:
a) KI   b) I_2 c) $NaIO_3$

2. Which of the reactions below can be classified as oxidation-reduction? For each one that is, identify the oxidizing agent and the reducing agent.

a) $2 Zn(s) + O_2 (s) \rightarrow 2 ZnO(s)$
b) $Zn(s) + I_2 (s) \rightarrow ZnI_2(s)$
c) $HI(aq) + NaOH(aq) \rightarrow NaI(aq) + H_2O$

3. One way to find the end point of a redox titration is to add a reagent that produces a color with an excess reactant. Can you think of another even simpler visual method that could be used for some reactions?

Instructor's Signature _____

Experiment 21
Redox Titration:
Iodine Used to Determine Ascorbic Acid

PURPOSE
Iodine solution from tincture of iodine will be standardized by titration with known ascorbic acid. The standard iodine can be used to titrate ascorbic acid products. Starch, that turns blue with iodine, is the indicator

INTRODUCTION
Ascorbic acid, or Vitamin C, is slowly oxidized by air. Unstable pharmaceutical products such as this must be closely monitored. Direct titration using iodine is a method that can be used to determine Vitamin C content in vitamin tablets and fruit juices.

Redox titration
Oxidizing agents gain electrons; reducing agents supply electrons. Ascorbic acid is easily oxidized and can be determined by titration with iodine, known as an iodimetric method. Iodine is the oxidizing agent in the reaction used to titrate ascorbic acid (the reducing agent):

Ascorbic acid   oxidized ascorbic acid

A known solution of ascorbic acid will be used to standardize the iodine solution. The standard iodine solution can then be used to analyze unknown ascorbic acid solutions prepared from fruit juices and commercial vitamin C tablets.

Since ascorbic acid begins to react with oxygen when dissolved in water, the solutions must be prepared within a few hours of their use.

Indicator
The starch molecule, amylose, is composed of glucose monomers twisted in a coiled structure. The channel within the spiral structure is thought to provide just enough space to accommodate an iodine/iodide species, forming a deep blue starch-iodine complex. In the titration of reducing agents such as ascorbic acid with iodine, the solution remains colorless until the end point when it turns dark blue in the presence of excess iodine. The end point is very close to the equivalence point.

APPARATUS

The titration apparatus includes a buret containing the standard iodine solution. The ascorbic acid sample is in the Erlenmeyer flask on a stir plate.

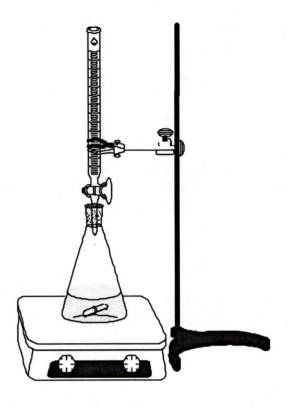

SAFETY

Wear safety glasses throughout this experiment.

Chemical	Toxicity	Flammability	Exposure
tincture of I_2	3	3	1
cornstarch	0	0	0
ascorbic acid	0	0	0

0 is low hazard, 3 is high hazard

PROCEDURE

The ascorbic acid will begin to oxidize once dissolved in water, and thus should not be stored more than a few hours before the titration is done. Iodine solution (from tincture of iodine) and starch indicator (from cornstarch) are provided.

Part A: Standardization of Iodine with Known Ascorbic Acid

1. Clean a 50-mL buret thoroughly with tap water, then rinse it with several small portions of the iodine solution, being sure to run some solution through the tip. Pour about 25 mL iodine solution in the buret. Fill the buret above the '0' mL mark, then lower the meniscus back to '0'.

2. Using a volumetric pipet and a pipet pump (or bulb) measure 25.0 mL of known ascorbic acid solution into a clean 125-mL Erlenmeyer flask. Record the molarity of the ascorbic acid.

3. Add about 1 mL (20 drops) of starch indicator solution.

4. Put a magnetic stir bar in the flask. Place the buret just inside the flask as shown in the figure so that the buret will drip directly into the solution.

5. Trial run: Perform a fast titration for an approximate end point. The appearance of the blue starch-iodine complex is the end point.

6. Careful run: Add iodine gradually (1 mL at a time) until you approach within 1 mL of the estimated endpoint. Then proceed drop by drop.

7. Perform another careful run with a second sample of the same ascorbic acid solution. If the careful runs are not very close to the same, see your instructor.

8. Find the concentration of the iodine solution.

Part B: Analysis of Vitamin C Tablets with Standard Iodine

1. Choose one of the Vitamin C tablets provided. Record the mass of ascorbic acid per tablet according to the product label. Crush the tablet and dissolve it in enough water to make a solution that is about 1 g/500 mL. Start by mixing the tablet with about 100 mL water in a beaker or Erlenmeyer. Then using a funnel, carefully transfer the solution into a volumetric flask. Be sure that all of the ascorbic acid is dissolved by washing any residue from the tablet with water. Fill the volumetric up to the mark.

2. Using a volumetric pipet and a pipet pump (or bulb) measure 25.0 mL of unknown ascorbic acid solution into a clean 125-mL Erlenmeyer flask.

3. Follow Steps 3 to 7 from Part A, only this time it is the concentration of the iodine solution that is known and the ascorbic acid that is unknown. Record the molarity of the standard iodine.

4. Find the concentration of the ascorbic acid solution.

5. Calculate the mass of ascorbic acid present in the sample titrated --- and the mass in an entire tablet. Compare with amount given on the product label.

Part C: Determination of Vitamin C in Fruit Juice

Give a procedure to find the mass of Vitamin C in fruit juice. Have the method checked by your instructor before proceeding. Record your results and compare with the product label. Note that labels give amounts of vitamin C as % RDA (recommended daily allowance) in an 8 oz serving (250 mL). For Vitamin C the RDA = 60 mg. So if cranberry juice contains 100% RDA it has 60 mg/250 mL. Clear juices, such as white cranberry juice, give the best results.

To estimate the volume of juice to titrate:

The amount of ascorbic acid titrated in Part A was a 25 mL aliquot of the solution that contained 1000 mg/500 mL. The sample titrated contained about 50 mg ascorbic acid and required about 18 mL iodine. You will need at least 30 mg of vitamin C in the juice sample in order to use more than 10 mL standard iodine solution. If possible, the volume of juice to be titrated should be less than 150 mL.

Data and Results (Redox Titration)

Name_____ Date_____ Section_____

Part A: Standardization of Iodine with Known Ascorbic Acid

Molarity of standard ascorbic acid _____ M

Titration	Equivalence Point (mL)
Trial run	
Careful run #1	
Careful run #2	
Average #1, #2	

Concentration of I_2 _____ M

Part B: Titration of Unknown Ascorbic Acid with Standard Iodine

Molarity of iodine: _____ M

Code number/letter of unknown ascorbic acid: _____

Titration	Equivalence Point (mL)
Trial run	
Careful run #1	
Careful run #2	
Average #1, #2	

Concentration of ascorbic acid _____ M

Instructor's Signature _____

Data and Results (Redox Titration)

Part C: Determination of Vitamin C in Fruit Juice

Questions

1. Why is it important to use at least 1.0 g ascorbic acid in preparing the standard solution?

2. A frozen lemonade product contains 15% RDA of Vitamin C in a prepared serving (water added). How large a sample of the prepared lemonade would you need so that at least 10 mL titrant would be required in a titration with iodine?

3. Why is an oxidation-reduction titration rather than acid-base titration used to detect ascorbic acid in juices?

4. Suppose a 25 mL sample of juice contained enough Vitamin C to require 10 mL standard iodine solution to reach the equivalence point. Calculate the RDA of Vitamin C for this juice. Does this seem to be a likely amount?

Experiment 22
Electrochemistry

Name_____ Date_____ Section_____

1. Assign oxidation numbers to the metallic element in each of the following:
a) $FeCl_2$ b) $FeCl_3$ c) $AlCl_3$ d) $CuSO_4$ e) Cu

2. Upon combining the standard half cells, Ni^{2+}(1 M)/Ni and Cu^{2+}(1 M)/Cu:
a) Which is the cathode and which is the anode?
b) Write the half-cell reactions and the overall cell reaction.
c) Describe the cell using conventional cell notation.
d) Find, $E°$, the standard cell potential.

3. In order to prepare 100 mL of 1 M $AlCl_3$, calculate the mass of Al metal that must be added to 1 M HCl. Assume 100% yield.

$$2Al(s) \ + \ 6HCl(aq) \rightarrow \ 2AlCl_3(aq) \ + \ 3H_2(g)$$

Instructor's Signature _____

Experiment 22
Electrochemistry

PURPOSE
Voltaic (galvanic) cells will be assembled and their potentials measured. An electrolytic cell is used to copper plate an object. Corroded metals will be cleaned using the principles involved in galvanic cells.

INTRODUCTION
Voltaic cells, electrolytic cells and the process of galvanic cleaning are described below.

Voltaic (Galvanic) Cell
In a voltaic cell an oxidation-reduction (redox) reaction produces electrical energy. One simple example is the Daniell cell made from copper and zinc half-cells. In the copper half-cell a strip of Cu metal is immersed in 1 M Cu^{2+} solution; in the zinc half-cell a Zn electrode is in contact with 1 M Zn^{2+} solution. Reduction or *gain* of electrons takes place at the copper cathode:

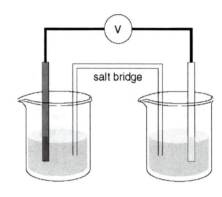

$$Cu^{2+} + 2\,e^- \rightarrow Cu$$

Oxidation or *loss* of electrons takes place at the zinc anode:

$$Zn - 2e^- \rightarrow Zn^{2+}$$

A salt bridge (shown as ||) containing concentrated electrolyte solution completes the circuit by connecting the half-cells. Using standard cell notation (cathode on the left and anode on the right), the Daniell cell is written:

$$Cu^{2+}(1M)|\ Cu(s)\ ||\ Zn(s)\ |\ Zn^{2+}\ (1M)$$

Upon combining standard reduction potentials (E°) the voltage for the copper/zinc cell is found to be 1.10 V (See Table 22.1).

Table 22.1 Standard Reduction Potentials

Half Cell Reaction	E°(V)
$Cu^{2+} + 2e^- \rightarrow Cu$	0.34
$Fe^{2+} + 2e^- \rightarrow Fe$	-0.45
$Zn^{2+} + 2e^- \rightarrow Zn$	-0.76
$Al^{3+} + 3e^- \rightarrow Al$	-1.66

Concentration Cell

In a concentration cell the half-cells are identical except for the concentration of the electrolyte solutions. The potential can be determined using the Nernst equation in which n is the number of electrons transferred in the cell reaction and Q depends on the concentration of the electrolyte solutions.

$$E = E° - \frac{0.059}{n} \log Q$$

For the cell reaction:

$$Zn(s) + Cu^{2+} = Zn^{2+} + Cu(s),$$

$$n = 2 \text{ and } Q = [Zn^{2+}]/[Cu^{2+}]$$

Electrolytic Cell

In an electrolytic cell electrical energy is supplied by a DC source (battery). The electrodes can be made of inert material since they are surfaces on which the redox reactions occur and do not participate in the reaction. When electrolysis is used for electroplating, the object to be plated is the cathode where reduction takes place. The solution contains the metal to be plated. For copper plating, Cu^{2+} ions in solution are reduced to copper metal.

$$Cu^{2+} + 2e^- \rightarrow Cu$$

Electrolysis is also used for decomposing compounds into their elements, for example NaCl into sodium metal and chlorine gas.

Galvanic Cleaning

For cleaning a metal, rusted iron or tarnished silver, for example, the corroded object is placed in a beaker of electrolyte and wrapped with a more reactive metal. The metal object is the cathode and the reactive metal is the anode. The electrolyte can be a solution of 10 percent sodium hydroxide (NaOH) or a 10 to 20 percent solution of the less caustic and safer to use sodium carbonate, (Na_2CO_3), available as washing soda. Baking soda, sodium bicarbonate $(NaHCO_3)$ is less effective. The process can be speeded up by heating. The object is left in the solution until the reactive metal completely oxidizes or the corrosion has been removed. Small, lightly corroded objects can be treated in this fashion.

To be effective, galvanic cleaning requires that sufficient metal core be present in the object being treated. Galvanic cleaning is impractical for large artifacts due to the amount of the anode that would be required. The method also obscures the object from view, making it difficult to observe the progress.

APPARATUS

A pH meter can be used as a voltmeter. The half-cells can be assembled in small beakers or even in the compartments of ice cube trays.

SAFETY
Wear safety glasses throughout the experiment.

Chemical	Toxicity	Flammability	Exposure
copper foil	0	0	0
iron metal	0	0	0
aluminum foil	0	0	0
1 M CuSO$_4$ · 5H$_2$O	2	0	2
1 M - 6 M HCl	2	0	2
1 M AlCl$_3$	1	0	1
1 M FeCl$_3$	1	0	1
1 M NaCl	0	0	0
10% Na$_2$CO$_3$	0	0	1

0 is low hazard, 3 is high hazard

PROCEDURE

Part A: Preparing Electrolyte Solutions
The Cu(II) solution has been prepared from CuSO$_4$ · 5H$_2$O.
The other two electrolyte solutions, Fe(II) and Al(III) can be made from reacting the metals with hydrochloric acid according to the procedures below. The concentrations will be roughly 1 M. The Fe(II) solution requires an overnight reaction and should be provided.

1. Pour 50 mL 3 M HCl into a 600-mL beaker. Add 3 g Al foil cut into small pieces. The reaction is vigorous, so add the pieces a few at a time.   Filter any unreacted metal. The Al(III) solution should be about 1 M.

2. Add 100 mL 6 M HCl to a 600-mL beaker. Add 20 g iron (staples) and allow to react overnight until the solution turns light green. Filter unreacted metal. The Fe(II) solution is about 3 M. Dilute to 150 mL to make approximately 1 M Fe(II) solution

Part B: Measuring Voltages Produced by Combining Half Cells

1. If you hold the red and the black leads together, the meter should read '0.0 mV'.

2. Check to see that the meter is operating correctly by measuring the voltage of a 1.5V battery.

3. Assemble the copper half-cell.   Fill one 30-mL beaker about 2/3 full with the blue 1 M copper sulfate solution. Fold a piece of the Cu foil over the beaker so that the foil is immersed in the solution.

4. Assemble the iron half-cell.   Fill another 30-mL beaker about 2/3 full with the light green 1 M Fe(II) solution. Scrape any oxide off the iron electrode (a nail, for example) by "scratching" the metal with the leads.

5. Attach the red lead to the Cu foil and the black lead to the Fe metal. Make a salt bridge by soaking a strip of filter paper in NaCl solution. Join the half-cells with the salt bridge. Record the measured voltage. Compare with the calculated voltage.

6. Repeat for Cu vs. Al. The aluminum will *always* be coated with oxide.

Part C: Measuring the Potential of a Concentration Cell

1. Dilute 1.00 mL of the 1 M $Cu(SO_4)_2$ to 0.001 M. (You can do this in two steps. First dilute the 1 M solution to 0.01 M by pipeting the 1 mL into a 100-mL volumetric flask, then filling to the mark with water. After careful mixing, pipet 1 mL from this 0.01 M solution into a 10-mL volumetric flask and fill with water.)

2. Prepare a concentration cell by preparing two half-cells, one from 1 M Cu^{2+} solution and the other with 0.001 M Cu^{2+}. Place a salt bridge across the two half-cells and attach the copper foils. Measure the cell potential.

3. Measure the potential produced by a different concentration cell using 1 M Cu^{2+} and 0.0001 M Cu^{2+} (instead of 0.001 M Cu^{2+}).

4. Compare the measured voltages with those predicted from the Nernst equation.

Part D: Electroplating

1. Fill a 30-mL beaker about half full with the 1 M copper sulfate solution.

2. Connect one lead from the cathode (object you want to plate) to the negative pole of the 9V battery. The other lead goes from the anode to the positive one. For the anode you can use a strip of copper metal or a penny dated before 1982. The cathode can be quarters, dimes, and paper clips, anything metallic. Nickels are smooth and somewhat difficult to plate. Immerse electrodes in the solution and observe what happens. If you remove the cathode from the solution you should see some copper deposited almost immediately. If not, check your connections.

3. Sketch and label the apparatus in the data/result section.

Part E: Galvanic Cleaning of Corroded Metals
Try using the principles involved in voltaic (galvanic) cells to clean corroded metal coins or nails provided. The materials you need are provided. Using sodium carbonate rather than sodium hydroxide is safer. Heating may be needed. Describe your method and record your observations.

Data and Results (Electrochemistry)

Name_____ Date_____ Section_____

Part B: Measuring Cell Potentials

	Voltage Measured (mV)	*Voltage Measured (V)	Cell
Cu vs. Fe			
Cu vs. Al			

* 1 V = 1000 mV

Part C: Measuring the Potential of a Concentration Cell

Half-cells		Voltage Meas'd (V)	Voltage Calc'd; Nernst Eq'n (V)
$Cu^{2+}($ M)	$Cu^{2+}($ M)		
$Cu^{2+}($ M)	$Cu^{2+}($ M)		

Part D: Electroplating

Sketch the setup used for electroplating, labeling the DC source, cathode, anode and electrolyte solution.

Instructor's Signature _____

Data and Results (Electrochemistry)

Part E: Galvanic Cleaning of Corroded Metals

Questions

1. Why is the voltage from a voltaic cell made from Cu vs. Al different from the value calculated using the reduction potentials in Table 22.1?

2. Make a labeled sketch of a set-up that could be used to silver plate an object.

3. Write the reaction that takes place at the cathode in the galvanic cleaning of a copper coin covered with oxide.

Experiment 23
Synthesis of Copper Nanoparticles

___

Name_____ Date_____ Section_____

1. For the length units:
i. centimeter   ii. Ångstrom   iii. micrometer
a) Give an example that can be used to illustrate the size of each one (such as an object or particle or distance).
b) Convert each unit to nanometers.

2. Compare a *colloid* consisting of a solid dispersed in water to a *solution* of a solid dissolved in water. Include properties and particle sizes.

3. Write half-reactions $(M^{+n} \rightarrow M^0)$ for the production of the following metals upon reduction of their salts:
a) silver from $AgNO_3$ b) copper from $CuCl_2$ c) gold from $AuCl_3$

Instructor's Signature _____

Experiment 23
Synthesis of Copper Nanoparticles

PURPOSE
Colloidal copper, containing copper nanoparticles (CuNPs) suspended in water, will be prepared by reduction of copper sulfate using glucose tablets.   Different brands of tablets will be compared.

INTRODUCTION
The size range and surprising color of noble metal nanoparticles is discussed as well as methods for their synthesis and identification.

Size and Color
Nanoparticles, which have dimensions between 1 and 100 nm, are compared to other "small" particles in the Figure below where you can see that the bacterium is huge in comparison.

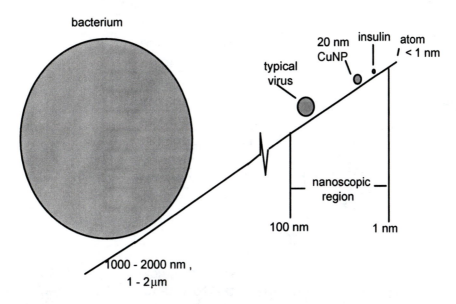

Properties of metal nanoparticles are different from those of bulk materials made from the same atoms. Color is the most striking example. The effect of nanoparticles on color has been known since antiquity when the tiny wine-red gold particles were used to make stained glass church windows.   In this experiment you will make copper nanoparticles in an aqueous medium. The color of the product will be different from the reddish brown of the bulk metal or the blue-green of aqueous copper ions.

Synthesis
Noble metal nanoparticles of gold, silver or copper can all be made by adding an excess of reducing agent to a salt of the metal. In this experiment copper sulfate pentahydrate provides Cu(II) ions. The reducing agent is glucose, $C_6H_{12}O_6$, which reduces copper ion

to elemental copper. The glucose itself is oxidized to produce gluconic acid, $C_6H_{12}O_7H$, which becomes the gluconate ion, $C_6H_{12}O_7^-$, in the presence of the base, NaOH.

The reaction can be summarized by the equation below (the solvent is distilled water):

$$C_6H_{12}O_6 \xrightarrow[Cu^{2+}/Cu^0]{OH^-} C_6H_{12}O_7^-$$

$$\text{glucose} \qquad\qquad\qquad \text{gluconate}$$

Adsorption of negative gluconate ions gives growing copper nanoparticles a negative surface charge. The nanoparticles are kept suspended in water separated from one another by repulsive electrostatic forces.

Glucose Tablets
Glucose tablets, taken to raise low blood sugar, contain ingredients in addition to glucose (Table 23.1) listed on the label in order from most to least. The identity of the ingredients depends on the brand of tablet.

Table 23.1 Ingredients in glucose tablets

Ingredients in Glucose Tablets	Formula	Solubility in H_2O
Ascorbic Acid	$C_6H_8O_6$	soluble
Carageenan	a polysaccharide	soluble
Citric Acid	$C_6H_8O_7$	soluble
Maltodextrin	$(C_{12}H_{20}O_{10})_n$	soluble
Xylitol	$C_5H_{12}O_5$	soluble
Yellow No. 6	$C_{16}H_{12}N_2O_7S_{2.2}Na$	soluble
Cellulose	$(C_6H_{10}O_5)_n$	insoluble
Magnesium Stearate	$C_{36}H_{70}MgO_4$	insoluble
Stearic Acid	$C_{18}H_{36}O_2$	insoluble

Identifying Copper Nanoparticles (CuNPs)
Nanoparticles are too small to be observed with an optical microscope. The image on the right was obtained with an electron microscope. The copper nanoparticles shown have diameters that are 18± 2 nm. (The length bar is 20 nm.) A 20-nm spherical particle contains about 150,000 Cu atoms. The appearance of color in the reaction mixture will be used to confirm the presence of CuNP's.

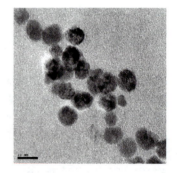

APPARATUS
Simple equipment including glassware and a hotplate is all that is needed for preparing colloidal copper --- copper nanoparticles suspended in water. In fact, this method for the synthesis of metal nanoparticles has been called "metallurgy in a beaker."

SAFETY

Wear safety glasses and gloves throughout this experiment. Copper sulfate stains the skin.

Chemical	Toxicity	Flammability	Exposure
glucose	0	0	0
1 M NaOH	2	0	2
0.010 M $CuSO_4 \cdot 5H_2O$	1	0	1

0 is low hazard, 3 is high hazard

PROCEDURE

Choose B.D. (Becton, Dickinson and Company) glucose tablets for Part A of this experiment. Enter the ingredients in the same order in which they appear on the label using the table in the Data/Results page.

Part A: Synthesis of Colloidal Copper

1. Calibrate a disposable pipet. Tare a small beaker, add 40 drops of water, and find the volume per drop (assuming the density of water is 1.0 g/mL).

2. Place 0.80 g glucose tablet in a 50-mL beaker. Using a graduated cylinder, add 10 mL distilled water. With a clean spatula, break up the tablet which should disintegrate and dissolve readily. Filter by gravity into another 50-mL beaker to remove insoluble substances such as magnesium stearate used to make the tablets.

3. Add two drops 1 M NaOH from a disposable pipet.

4. Heat to 50°C on a hotplate. Determine the temperature by immersing a thermometer in a second 50-mL beaker containing water. The thermometer should not touch the bottom of the beaker.

5. Add 0.20 mL 0.010 M $CuSO_4 \cdot 5H_2O$ by counting drops using the calibrated disposable pipet. Heat until a yellow color appears. Remove from heat. Record the time required for color to appear.

6. Cool to room temperature. Describe the product. Observe any changes in appearance with time.

Part B: Substituting Other Brands of Glucose Tablets

1. Choose another brand of glucose tablet other than the B. D. tablets used in Part A. Record the ingredients in the order in which they appear on the label.

2. Repeat the synthesis from Part A and record results.

3. Combining your result with those of other groups complete the table for all of the brands tested in Section B.

4. Which ingredient or ingredients could be inhibiting the reaction?

Data and Results (Copper Nanoparticles)

Name_____ Date_____ Section_____

Part A: Synthesis of Colloidal Copper Using B.D. Glucose Tablets

Describe the reaction including the appearance of the product.

Part B: Synthesis of Colloidal Copper Using Other Glucose Tablets

Brand of Glucose Tablet	Ingredients (in the same order in which they are listed on the label)	Result of Reaction with Cu(II)
B.D.		

Instructor's Signature _____

Questions

1. For the reaction used in this experiment, give the half reaction for the reduction of ionic copper, and show that copper ion is the oxidizing agent.

2. Which reactant, copper sulfate pentahydrate or glucose, is in excess (the reaction stoichiometry is one mol glucose to one mol copper salt)? How does the excess reactant stabilize the copper nanoparticles?

3. Show that the number of copper atoms in a 20-nm copper nanoparticle is about 150,000. Assume that each copper atom occupies the volume of a cube with an edge of 0.3 nm. (Volume of a sphere = $4/3 \, \pi \, r^3$)

4. Which brands (or brand) of glucose tablets produced a color change during the reaction? Which did not? From examining results contributed by the entire class, determine what ingredient or ingredients could be responsible for this. Give your reasoning.

Experiment 24
Synthesis of Copper Pigments: Making Paints

Name_____ Date_____ Section_____

1. The pigment, chrome yellow, is prepared by mixing solutions of lead(II) nitrate and sodium chromate.
a) Write a balanced chemical equation for this reaction.
b) What is the formula of chrome yellow?

2. Another yellow pigment is made from mixing solutions of $Cd(NO)_3 \cdot 4H_2O$ and $Na_2S \cdot 9H_2O$. The resulting precipitate is filtered and dried.
a) Write a balanced chemical equation for this reaction.
b) What is the formula of cadmium yellow?
c) Suppose a 50-mL solution containing 9.7 g hydrated cadmium nitrate is combined with a 50-mL solution containing 8.3 g hydrated sodium sulfide. Which is the excess reactant?

3. Vermilion, synthetic red mercuric sulfide, can be made from mixing and heating elemental mercury and sulfur using a process known before 800 AD, according to the reaction:

$$Hg + S \rightarrow HgS$$

a) If 30 g Hg is mixed with excess S to produce 31 g HgS, what is the % yield?
b) What problems could arise in attempting to make vermilion in a general chemistry laboratory?

Instructor's Signature _____

Data and Results (Pigments)

Name_____ Date_____ Section_____

Part A: Synthesis of Malachite

Mass $CuSO_4 \cdot 5H_2O$ _____ g

Moles $CuSO_4 \cdot 5H_2O$ _____ mol

Mass Na_2CO_3 _____ g

Moles Na_2CO_3 _____ mol

Limiting Reactant _____

Theoretical mass $Cu_2(OH)_2CO_3$ _____ g

Actual mass $Cu_2(OH)_2CO_3$ _____ g

Yield $Cu_2(OH)_2CO_3$, malachite,_____ %

Instructor's Signature _____

Data and Results (Pigments)

Part B: Making Verdigris, Copper Acetate

Mass $CuSO_4 \cdot 5H_2O$ _____ g   Moles $CuSO_4 \cdot 5H_2O$ _____ mol

Moles $Cu(OH)_2$ _____ mol

Theoretical mass $Cu(CH_3COO)_2 \cdot H_2O$ _____ g

Actual mass $Cu(CH_3COO)_2 \cdot H_2O$ _____ g

Yield $Cu(CH_3COO)_2 \cdot H_2O$ _____ %

Part C: Making and Testing Paints

Questions

1. Why is it particularly important to wash the cupric hydroxide intermediate in the synthesis of cupric acetate (verdigris)? Be specific about what is being removed and why it matters.

2. Why is it more difficult to precipitate cupric acetate (verdigris) than cupric carbonate hydroxide (malachite)?

3. Could blue copper sulfate pentahydrate be ground and used as a pigment for making paint? Why?

4. What feature is shared by formulas of colored compounds used as pigments?

Experiment 25
Synthesis of Ethyl Salicylate from Aspirin Tablets

Name_____ Date_____ Section_____

1. In the equilibrium reaction below a carboxylic acid reacts with an alcohol to produce an ester and water.

$$R-\overset{\overset{\textstyle O}{\|}}{C}-OH \quad + \quad HOR' \quad \rightleftharpoons \quad R-\overset{\overset{\textstyle O}{\|}}{C}-OR' \quad + \quad H_2O$$

a) Write the equilibrium constant, K_{eq}, for the reaction.
b) Using LeChatlier's Principle list some ways by which the reaction could be driven to the right, toward the ester product.

2. Name the esters below:

a) $CH_3-\overset{\overset{\textstyle O}{\|}}{C}-OCH_3$ b) $CH_3-\overset{\overset{\textstyle O}{\|}}{C}-OCH_2CH_3$ c) ⟨benzene ring⟩$-\overset{\overset{\textstyle O}{\|}}{C}-OCH_2CH_3$

3. Consider the reaction scheme in which A is first converted to B and then B to C. If the yield is 91% for reaction I and 75 % for reaction II, what is the overall yield of the formation of C?

I	A	→	B
II	B	→	C
III	A	→	C

Instructor's Signature _____

Prelaboratory Exercise

Experiment 26

Qualitative Analysis of Household Compounds and Mixtures

Name________________________________ Date______________ Section______

1. The compounds below can all be classified as soluble or insoluble in water. For each compound, give its name and decide whether it is water-soluble. When possible, add everyday observations to support your decisions about solubility.

a) $NaCl$	b) $CaCO_3$	c) $NaHCO_3$	d) KI
e) $CaSO_4$	f) $CuCO_3$	g) $Cu(OH)_2$	h) CH_3CH_2OH

2. Of the liquids or solutions below, which are acidic? alkaline? neutral? (HAc is acetic acid, $HC_2H_3O_2$)

a) $NaOH(aq)$	b) $HCl(aq)$	c) $NH_3(aq)$	d) $KCl(aq)$
e) H_2O (l)	f) $Na_2CO_3(aq)$	g) $NH_4Cl(aq)$	h) $HAc(aq)$

3. Complete and balance the following chemical equations:

a) $NaOH(aq) + HAc(aq) \rightarrow$

b) $CaCO_3 + HCl(aq) \rightarrow$

c) $Al(s) + HCl(aq) \rightarrow$

d) $Mg(OH)_2(s) =$

*Instructor's Signature*______________________________

Experiment 27
Measuring Potassium-40 in Salt Products

Name_____ Date_____ Section_____

1. There are three naturally occurring isotopes of potassium. Using the abundance and atomic mass information given, calculate the atomic weight of the element, K.

Isotope	abundance %	atomic mass
K-39	93.258	38.964
K-40	0.012	39.964
K-41	6.730	40.962

2. The amount of potassium in the body is about 140 g. How much of this is K-40? (Use data in Question 1).

3. The label on a salt substitute product containing KCl says that 1 g provides 530 mg potassium. Is this possible? Explain.

Instructor's Signature _____

Experiment 27
Measuring Potassium-40 in Salt Products

PURPOSE
The amount of potassium in salt products will be found by measuring K-40 content and comparing it to a standard potassium source. This experiment can also be done as a dry lab. Data will be provided for analysis.

INTRODUCTION
The spontaneous decay of radioactive isotopes is followed by a discussion of the sources and uses of Potassium-40.

Spontaneous Decay
Radioactive elements emit alpha, beta or gamma radiation to become a new element or a new isotope of the same element. Alpha particles, α, are helium nuclei which have a charge of +2 and a mass number of 4. Beta minus particles, β, are electrons with a charge of -1 and a mass number of 0. Gamma, γ, is electromagnetic radiation similar to x-radiation, and thus has 0 mass and 0 charge. The 'm' in isotopes which decay by emitting γ, such as Tc-99m, means metastable. Examples of nuclear equations for each type of decay mode are given below. Notice that the masses and charges balance. The identity of a new element formed is found from its charge, which is the atomic number of the element. Only heavier isotopes emit alpha. For instance, U-238 emits alpha particles and forms thorium-234.

$$^{238}_{92}U \rightarrow ^{234}_{90}Th + ^{4}_{2}\alpha$$

The short-lived gamma emitter, Tc-99m, with a half-life of about 6 hours, is used as a diagnostic tool in nuclear medicine. The newly formed element is still Tc-99 but is no longer metastable.

$$^{99m}_{43}Tc \rightarrow ^{99}_{43}Tc + ^{0}_{0}\gamma$$

An example of a beta emitter is K-40 that decays to produce Ca-40. A second route by which K-40 decays is known as electron capture or EC, this time producing an isotope of argon, Ar-40:

$$^{40}_{19}K \rightarrow ^{40}_{20}Ca + ^{0}_{-1}\beta \qquad ^{40}_{19}K + ^{0}_{-1}e \rightarrow ^{40}_{18}Ar + ^{0}_{-1}\beta$$

Of the decay modes of K-40, 89.3% is beta emission and the remaining 10.7% is by electron capture.

Potassium-40

Wherever there is K, there is also K-40. The isotope K-40 accounts for 0.012% of all potassium atoms and is the predominant radioactive isotope in human tissues and in most foods. The human body controls the level of K to maintain the normal level required for systems to function. Therefore, consuming large amounts of foods containing the element, such as salt substitutes or bananas, will not increase the amount of K (or K-40) in the body.

Substitutes for NaCl, required by people on low sodium diets, contain varying amounts of KCl. One product that is nearly 100% KCl will be used as the standard K-40 source for this experiment. The amount of K-40 in other salts containing mixtures of NaCl and KCl will be determined by comparison with the standard source.

APPARATUS

A radiation monitor (Geiger counter) connected to a computer through the Vernier LabPro panel is used to monitor radiation.

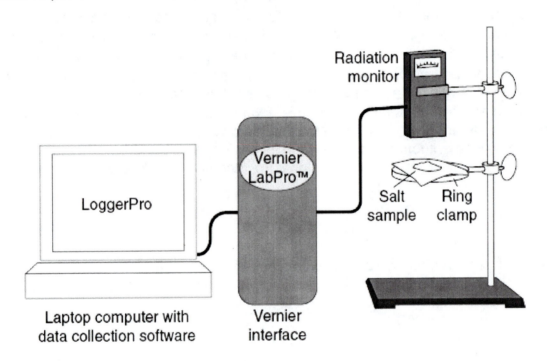

The Geiger counter contains a tube of argon gas, Ar, and a thin entrance window for the radiation. Radiation entering the window ionizes the Ar forming an "ion pair" which consists of an argon cation, Ar^+, and an electron. When ion pairs are produced an electric circuit is completed and counts can be detected. The number of counts per minute, cpm, measures the quantity of radioactivity. The positioning of the sample is important since radiation drops off with increased distance.   If the distance from a point source of radiation is doubled, the exposure is only 1/4 of that at the original distance.

SAFETY

Wear safety glasses throughout this experiment.

Chemical	Toxicity	Flammability	Exposure
sodium chloride	0	0	0
potassium chloride	0	0	0

0 is low hazard, 3 is high hazard

PROCEDURE

Part A: Setting Up the Computer-Interfaced Probe

1. Plug in the Vernier interface to an outlet. You will hear a series of beeps. Connect the radiation monitor to the Dig/Sonic 1 (DG 1) port of the Vernier interface. Attach the interface to the back of the computer using the USB cord.

2. Attach the radiation monitor to the ring clamp on the ring stand using a two-pronged test-tube clamp. (See apparatus). Turn on the radiation monitor. Set the range switch to BATT. If the meter pointer doesn't indicate within the battery section on the scale, notify your instructor. (The meter takes a common 9 V battery.) After checking the battery, set the range switch to the x1 position and set the function switch to AUDIO.

3. Open the LoggerPro software application on the computer desktop. Click on the LabPro icon above the graph window to open the Sensors window. Drag the Radiation RM-BTD icon from the left hand side of the Sensors window to the DIG/SONIC1 port of the display of the LabPro interface. Close the Sensors window. Click on the clock icon next to the LabPro icon. The Data Collection Screen appears.
   • Mode is preset to time-based
   • Length [enter 60 and min]
   • Sampling Rate: [enter 0.2/min]
                          5 min intervals
The computer will display (and plot) columns of data showing total counts for each 5 min interval, such as:
      5   81
    10 70 and so on
That means there were 81 counts in the first 5-min interval and 70 counts in the next 5-min interval, and so on. Total counts were rounded to the nearest count.

Part B: Measuring Background Radiation

1. Click on the **Collect** button located above the chart section on the computer monitor. Data will automatically be entered into the **Latest** window and plotted on the chart. (Note that the data is sampled every five minutes.) For total counts add the 12 readings from the computer table, then divide by 60 to obtain counts per minute (cpm). This value will be subtracted from future sample readings.

2. The **Collect** button will reappear.

Part C: Radiation Level of Salt Products

1. Record the mass of salt product to be used. Position the sample 1.0 cm from the radiation window by moving the ring clamp as needed.

2. Click on the **Collect** button located above the chart section on the computer monitor. Data will automatically be entered into the **Latest** window and plotted on the chart. Collect the data for 60 min and take a total of the readings and divide the total by 60 to obtain counts per minute (cpm).

3. Subtract the background reading from this value and record the result in the data table.

4. Repeat with other salts, again positioning the radiation monitor so that the mica window is focused on the middle of the salt bottle at a distance of 1.0 cm.

5. Compare cpm and amount of potassium for all salt samples, subtracting background each time.

Data and Results (Nuclear Chemistry)

Name_____ Date _____ Sec._____

Part B: Measuring Background Radiation

Time Interval											
Counts/ Interval											

Total counts _____

Total counts/ total minutes _____

Part C: Radiation Level of Salt Products

Sample _____ Mass _____ g

Time Interval											
Counts/ Interval											

Total counts _____

Total counts/ total minutes _____

Sample _____ Mass _____ g

Time Interval											
Counts/ Interval											

Total counts _____

Total counts/ total minutes _____

Instructor's Signature _____

Data and Results (Nuclear Chemistry)

Sample _____   Mass _____ g

Time Interval											
Counts/ Interval											

Total counts _____

Total counts/ total minutes _____

Sample _____   Mass _____ g

Time Interval											
Counts/ Interval											

Total counts _____

Total counts/ total minutes _____

Sample _____   Mass _____ g

Time Interval											
Counts/ Interval											

Total counts _____

Total counts/ total minutes _____

Questions

1. One medium banana, a food that is rich in potassium, weighs about 150 g and contains about 420 mg potassium. Comment on whether the K content of a banana could be found by measuring its K-40 content as done for the salts used in this experiment. Describe any problems that might be encountered.

2. Compare your cpm results to the amount of potassium on the product label. Why is it all right to use g K rather than atoms K, g K-40 or atoms K-40?

Sample	cpm	g K (from label)

3. What are the errors that could arise in this experiment?

4. Would it be easier or more difficult to estimate the K-40 and thus K, content of a salt substitute consisting of KI as the major ingredient? Explain.

Appendix A Table of Available Chemicals

The tables in this appendix list household chemicals that are readily available including all of those from experiments in this manual. The information tabulated includes the name, formula, molar mass, melting or boiling point, density, solubility in water and source for each chemical.

Densities of all solids are at room temperature (near 20°C). For liquid densities the temperature is specified and for gas densities, both temperature and pressure.

Solubilities in water are at room temperature unless otherwise noted. When numerical solubilities are not available, abbreviations including those in Table A.1 are used instead.

Table A.1

Abbreviation	Meaning	g/100g Water
∞	Infinitely soluble; miscible	unlimited
vs	very soluble	30 – 100's
s	soluble	1 - 30
sl s	slightly soluble	0.001 - 0.01
i	insoluble	< 0.001

Melting points and *boiling points* are listed in the same column; the boiling points are in parentheses.

Sources given are either products that contain the chemical or a method by which the chemical can be prepared using liquid extraction or synthesis.

Table A.2

Name	Formula	Molar Mass	MP (BP) (°C)	Density (g/mL)	Sol H$_2$O (g/100g)	Source
acetaminophen	C$_8$H$_9$NO$_2$	151	169-170.5	1.293	1.4	Tylenol®
acetamide	C$_2$H$_5$NO	59	81	1.16	200	HAc $_{(aq)}$ + NH$_{3\,(aq)}$ + Heat
acetic acid (aq)	C$_2$H$_4$O$_{2\,(aq)}$	60				vinegar, 5% (aq)
acetone	C$_3$H$_6$O	58	(56.5)	0.788 @25°C	∞	nail polish remover
acetyl salicylic acid	C$_9$H$_8$O$_4$	180	135	1.40	4.6	aspirin tablets
light alkanes (n= 5-7)	C$_n$H$_{2n+2}$	72 - 100	(36 - 98)	0.63 - 0.68	i	lighter fluid
medium alkanes (n= 8-10)	C$_n$H$_{2n+2}$	114 - 142	(125 - 173)	0.70 - 0.73	i	charcoal lighter adhesive remover
heavy alkanes (n= 11-13)	C$_n$H$_{2n+2}$	156 - 184	(194 - 234)	0.74 - 0.75	i	adhesive remover lamp oil
aluminum	Al	27	660	2.70	i	foil
aluminum chloride	AlCl$_3$	133.34	178 sublimes	2.44	45.8	Al + HCl $_{(aq)}$
ammonia (aq)	NH$_{3\,(aq)}$	17	-	10% 0.96 @25°C		3 -10% in cleaning products
ammonium acetate	C$_2$H$_7$NO$_2$	77	114	1.07	150	HAc $_{(aq)}$ + NH$_{3\,(aq)}$
ammonium bicarbonate	NH$_4$HCO$_3$	79	106	1.586	17.4	NaHCO$_3$ + NH$_{3\,(aq)}$
ammonium chloride	NH$_4$Cl	53.5	sub	1.527	27	NH$_{3\,(aq)}$ + HCl $_{(aq)}$
iso-amyl acetate	C$_7$H$_{14}$O$_2$	130	(142)	0.576 @15°C	1.6	nail polish remover
anthocyanin						red cabbage
ascorbic acid	C$_6$H$_8$O$_6$	176	190	1.65	33	vitamin C
beeswax	-	-	62 - 65	0.95-0.96	i	candles
boric acid	H$_3$BO$_3$	61.8	171	2.46	5	bug killer
butane	C$_4$H$_{10}$	58	(-0.5)	2.4g/L @ RT,1atm	0.0061	lighter fluid
caffeine	C$_8$H$_{10}$N$_4$O$_2$	194	238	1.23	2.17	Caffedrine capsules No-Doz tablets

Table A.3

Name	Formula	Molar Mass	MP (BP) (°C)	Density (g/mL)	Sol H$_2$O (g/100g)	Source
calcium carbonate	CaCO$_3$	100	825	2.83	i	chalk
calcium chloride anh.	CaCl$_2$	111	772	2.15	74.5	ice melting product
calcium oxide	CaO	56	2572	3.3	5	lime
calcium sulfate hemihydrate	CaSO$_4$· 0.5H$_2$O	136	1460	2.96	0.2	Plaster of Paris
camphor	C$_{10}$H$_{16}$O	152	179	0.992	0.12	
carbon (graphite)	C	12	4027-4427	2.267		pencil lead
catalase	protein	240 x 10^3				yeast, potatoes
cellulose	~C$_6$H$_{10}$O$_5$~			0.600	i	cotton, paper, wood
chlorophyll a	C$_{55}$H$_{72}$Mg N$_4$O$_5$	893.49	117-120		i	spinach
chlorophyll b	C$_{55}$H$_{70}$Mg N$_4$O$_6$	907.47	183-185		i	spinach
citric acid	C$_6$H$_8$O$_7$	192	153	1.665	59.2	lemons
copper	Cu	63.55	1083	8.94		wire, pennies (pre1982)
copper sulfate pentahydrate	CuSO$_4$· 5H$_2$O	249.68	650 dec.	2.286	about 35	root killer
starch	~C$_6$H$_{10}$O$_5$~					cornstarch
curcumin	C$_{21}$H$_{20}$O$_6$	368.4	183		i	turmeric
dextrose (see D-glucose)						
p-dichloro-benzene	C$_6$H$_2$Cl$_2$	147	53.5	1.46	i	mothballs
ethanol	C$_2$H$_6$O	46	(78.5)	0.7893	∞	rubbing alcohol
ethyl acetate	C$_4$H$_8$O$_2$	88	(77)	0.902@20 0.898@25	10	nail polish remover
ethyl salicylate	C$_9$H$_{10}$O$_3$	166	(231-234)	1.131	slightly	salicylic acid + EtOH
ethyl vanillin	C$_9$H$_{10}$O$_3$	166	77-78		slightly	Tone's imitation vanilla

Table A.4

Name	Formula	Molar Mass	MP (BP) (°C)	Density (g/mL)	Sol H$_2$O (g/100g)	Source
ethylene glycol	C$_2$H$_6$O$_2$	62	-13 (198)	1.1135	∞	antifreeze
fructose	C$_6$H$_{12}$O$_6$	180	103-105		400	fruit sugar
gelatin	protein					Knox® gelatin
D-glucose	C$_6$H$_{12}$O$_6$	180	146-152	1.54	90	glucose tablets
glutamic acid	C$_5$H$_9$NO$_4$	147	247 -249	1.54	0.86	monosodium glutamate
glycerol	C$_3$H$_8$O$_3$	92	(290)	1.264@20 1.262@25	∞	
hydrochloric acid (aq)	HCl $_{(aq)}$	36.5		20% 1.10@20°		20% and 32% muriatic acid
hydrogen peroxide (aq)	H$_2$O$_2$ $_{(aq)}$	34		3% about 1.00		3% aqueous
iodine	I$_2$	254	113.6 (184.2)	4.93@25 3.96@20		2% I$_2$ (in tincture)
iron	Fe	55.85	1535	7.86	i	steel wool, nails
iron(II) chloride hexahydrate	FeCl$_2$· 6H$_2$O	199		1.93	About 70	xsFe + HCl $_{(aq)}$
iron(III) chloride tetrahydrate	FeCl$_3$· 4H$_2$O	270		2.90	92	Fe + xsHCl $_{(aq)}$
isopropyl alcohol	C$_3$H$_8$O	60	(82.4)	0.840 @20°C	∞	rubbing alcohol
limonene	C$_{10}$H$_{16}$	136	(175.5-176.5)	1.184 @20°C	i	oils of lemon and orange
magnesium hydroxide	Mg(OH)$_2$	58	350 dec.	2.36	0.0012	MgSO$_4$ + NaOH
magnesium sulfate hydrate	MgSO$_4$· 7H$_2$O	246	1124 dec.	1.67	125	Epsom salt
menthol	C$_{10}$H$_{20}$O	156	41-43 (212)	0.890	slightly	peppermint oil
methanol	CH$_4$O	32	(64.7)	0.791@20 0.787@25	∞	antifreeze, canned heat, octane booster
methyl ethyl ketone	C$_4$H$_8$O	72	(79.6)	0.805 @20°C	12.5	cleaning solvent
methyl salicylate	C$_8$H$_8$O$_3$	152	(220-224)	1.184 @25°C	0.07	oil of wintergreen
monosodium glutamate	C$_5$H$_8$NNaO$_4$	169	225	26.2	41.7	seasoning

Table A.5

Name	Formula	Molar Mass	MP (BP) (°C)	Density (g/mL)	Sol H$_2$O (g/100g)	Source
naphthalene	C$_{10}$H$_8$	128	80.2	1.162	i	moth balls
oxygen (g)	O$_2$	32		1.33 g/L RT,1 atm		H$_2$O$_2$ + catalase
papain	protein	about 23,400			i	meat tenderizer
paraffin wax	C$_n$H$_{2n+2}$	-	50-57	0.90	i	candles
phenol, carbolic acid	C$_6$H$_6$O	94	43	1.071	6.67	
polyethylene	~CH$_2$CH$_2$~	1,5-100 x 10^3	85-110	0.92	i	plastic wrap
polystyrene	~CHCH(Ph)~	-	about 85	1.04-1.065	i	plastic film, cups
polyvinyl chloride	~CH$_2$CHCl~	60-150 x 10^3	212	1.406	i	
potassium acid tartrate	C$_4$H$_4$K$_2$O$_6$	226	150	1.98	0.5	cream of tartar
potassium chloride	KCl	74.55	773	1.98	35.7	salt substitute
potassium iodide	KI	166	680	3.12	140	tincture iodide
salicylic acid	C$_7$H$_6$O$_3$	138	157-59	1.443	0.20	aspirin + HCl
sodium acetate trihydrate	C$_2$H$_3$O$_2$Na· 3H$_2$O	136	58	1.45	125	HAc $_{(aq)}$ + NaHCO$_3$
sodium bicarbonate	NaHCO$_3$	84	270	2.159	10	baking soda
sodium borate decahydrate	Na$_2$B$_4$O$_7$· 10H$_2$O	381		1.73	6.3	Borax
sodium carbonate	Na$_2$CO$_3$	106	851	2.532	30	Arm & Hammer® washing soda
sodium chloride	NaCl	58.5	804	2.17	35.7	table salt
sodium hydroxide	NaOH	40	318	2.13	111	lye; oven cleaner
sodium hypochlorite (aq)	NaOCl $_{(aq)}$	74.44				5% in bleach
sodium salicylate	C$_7$H$_5$NaO$_3$	160	440		125	salicylic acid + NaOH
sodium stearate	C$_{18}$H$_{35}$NaO$_2$	306	205	1.02	slightly	soap

Table A.6

Name	Formula	Molar Mass	MP (BP) (°C)	Density (g/mL)	Sol H_2O (g/100g)	Source
sodium stearate	$C_{18}H_{35}NaO_2$	306	205	1.02	slightly	soap
sodium thiosulfate	$Na_2O_3S_2$	158	48	1.69		
stearic acid	$C_{18}H_{36}O_2$	284.5	69-70	0.85	i	
sucrose	$C_{12}H_{22}O_{11}$	342	185-187 dec.	1.587	211.5	table sugar
tannic acid	$C_{76}H_{52}O_{46}$	1701	210-215		285	acorns, tea
L-tartaric acid	$C_4H_6O_6$	150	168-170	1.76	133	from potassium bitartrate
trisodium phosphate	Na_3PO_4	164	75	1.6	3.5	
tungsten	W	183.84	3410	18.7-19.3	i	light bulbs
urea	CH_4N_2O	60	132.7	1.32	108	fertilizer
vanillin	$C_8H_8O_3$	152	81-83	1.056	2	pure vanilla extract, beans
zinc	Zn	65.39		7.14	i	batteries

CPSIA information can be obtained at www.ICGtesting.com
Printed in the USA
BVOW06n1558031214

377619BV00002B/2/P